THE BMW CENTURY

THE **ULTIMATE PERFORMANCE** MACHINES

THE BMW CENTURY

THE **ULTIMATE PERFORMANCE** MACHINES

TONY LEWIN

Foreword by Tom Purves

m
motorbooks

Brimming with creative inspiration, how-to projects, and useful information to enrich your everyday life, Quarto Knows is a favorite destination for those pursuing their interests and passions. Visit our site and dig deeper with our books into your area of interest: Quarto Creates, Quarto Cooks, Quarto Homes, Quarto Lives, Quarto Drives, Quarto Explores, Quarto Gifts, or Quarto Kids.

© 2016 Quarto Publishing Group USA Inc.
Text © 2016 Tony Lewin

First published in 2016 by Motorbooks, an imprint of The Quarto Group, 401 Second Avenue North, Suite 310, Minneapolis, MN 55401 USA. T (612) 344-8100 F (612) 344-8692 www.QuartoKnows.com

Motorbooks titles are also available at discount for retail, wholesale, promotional, and bulk purchase. For details, contact the Special Sales Manager by email at specialsales@quarto.com or by mail at The Quarto Group, Attn: Special Sales Manager, 401 Second Avenue North, Suite 310, Minneapolis, MN 55401 USA.

10 9 8 7 6 5 4 3

ISBN: 978-0-7603-5017-1

Library of Congress Cataloging-in-Publication Data

Names: Lewin, Tony.
Title: The BMW century : the ultimate performance machines / Tony Lewin.
Description: Minneapolis, Minnesota : Quarto Publishing Group Inc.,
    Motorbooks, 2016.
Identifiers: LCCN 2016015442 | ISBN 9780760350171 (hardback)
Subjects: LCSH: BMW automobiles--History. | BMW motorcycle--History. | BISAC:
    TRANSPORTATION / Automotive / History. | TRANSPORTATION / Automotive /
    Pictorial. | TECHNOLOGY & ENGINEERING / Automotive.
Classification: LCC TL215.B25 L493 2016 | DDC 338.7/629222--dc23
LC record available at https://lccn.loc.gov/2016015442

Acquiring Editor: Zack Miller
Project Manager: Madeleine Vasaly
Art Director: Brad Springer
Cover Designer: Jay Smith–Juicebox Designs
Interior Design: Brad Norr
Layout: Simon Larkin

All images courtesy BMW AG PressClub and BMW Archive except as noted otherwise.

On the front cover: A close view of the badge on BMW's i8 hybrid sports car. *William Stern*
On the front endpaper: BMW's celebrated 507 poses on a winter mountain pass.
On the rear endpapers: BMW's portfolio of Hommage concept models and the classic cars
    and bikes that inspired them. In the foreground is the 2002 Hommage, revealed in May 2016.
On the title page: BMW's Berlin dealership in 1929. Clearly visible are the just-launched
    3/15 sedan and the R62 motorcycle, complete with Steib sidecar.

Printed in China

Contents

Foreword

By Tom Purves, Chairman of the Royal Automobile Club, Pall Mall, London

For twenty-five years I worked for BMW, first as sales director and then as managing director of BMW (GB) Ltd. in the 1980s and 1990s. Following this I was sales director of the Rover Group during the period of BMW ownership. I then spent nearly ten years as president of BMW North America and chairman of BMW US Holding Corporation before completing my professional career within the group as chief executive of Rolls-Royce Motor Cars Ltd. at Goodwood. As I had started my life in the automobile industry, forty-three years earlier, as a student engineering apprentice at Rolls-Royce's car division at Crewe, it was a fitting and fulfilling conclusion.

BMW is a company run by engineers. As an engineer turned sales and marketing man, I revelled in never having to justify poor design, poor development, or poor manufacturing: whilst I was always aware of the relatively low esteem in which sales people were once held at BMW, I considered that a price worth paying for an exemplary product to sell.

At its happiest as an engine maker—think of the power of the early radial aero engines, the torque of the boxer motorcycle engine, or the silky smoothness of the six-cylinder petrol engines that did so much to characterise the sports sedans of the post–Second World War era—BMW was innovative in all areas. The light weight of the 328 sports car in the 1930s, the early use of digital engine management systems in the 1970s, and the clever front fork design and pioneering ABS for the motorcycles of the 1980s—all these are examples of BMW's flair for innovation.

However, when Tony Lewin asked me to write the foreword to his book, I reflected on the single most important element which has given the company its enormous success in modern times. I worked under the group chairmanship of Eberhard von Kuenheim, Bernd Pischetsrieder, Joachim Milberg, Helmut Panke, and Norbert Reithofer. The single focus each one gave to the long-term success of the company and their clear insistence that no one, least of all the top man, was more important than the company, contributed more than anything else to the achievements of the organisation. The atmosphere in management ranks was never complacent: it was always demanding, and, like the engines powering the products, it was certainly highly efficient. These strong leaders, who never forgot they were directing a car company, all held the belief that if you built the best products, people would buy them. And this is how BMW has achieved its remarkable success.

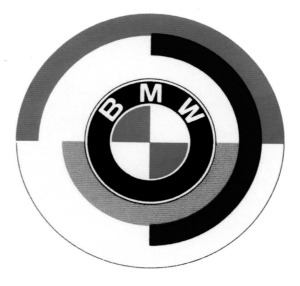

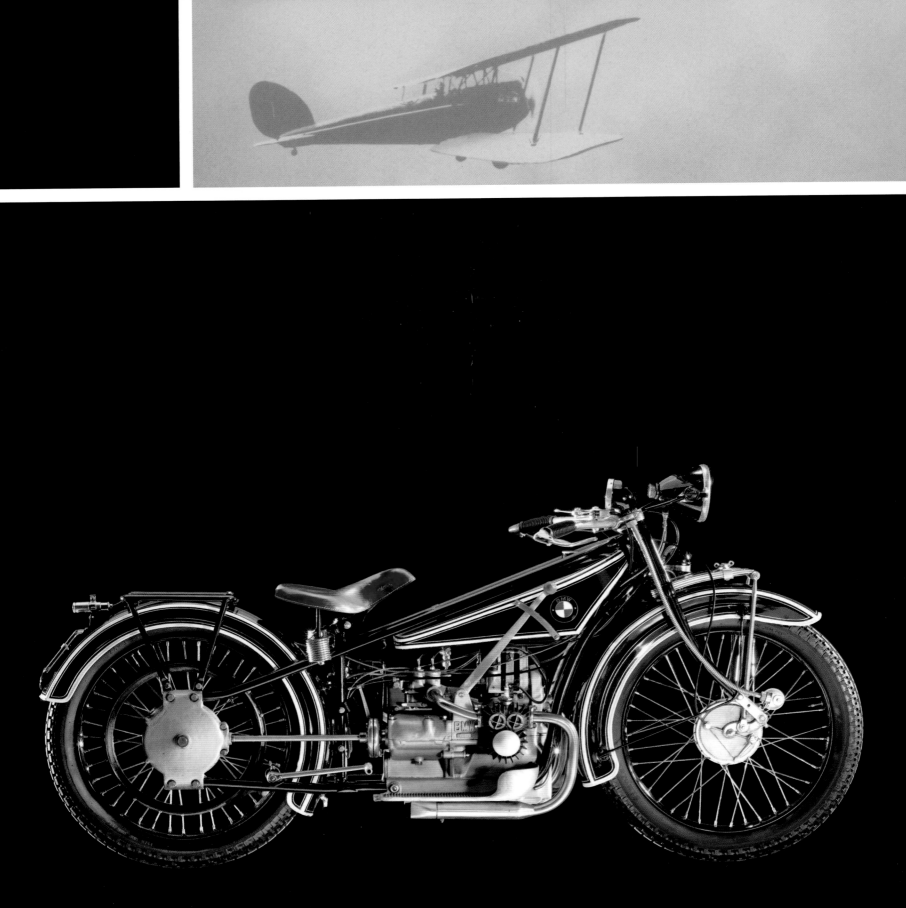

1

High Flier
from the Start

BMW has always liked to do things differently—and this held true even before its constituent parts took on the BMW name or began using the blue-and-white roundel badge.

Most of the successful carmakers of the early twentieth century could point—often with great pride—to origins that exactly paralleled the growing desire for greater mobility among their customers. Typically, these resourceful operators emerged in the late 1800s, first making bicycles, then moving on to crude motor-assisted cycles, and then producing motorcycles, multiwheeled cyclecars, lightweight cars, and finally proper cars with decent engines, steering, and brakes. Others, such as BSA, Škoda, and Hotchkiss, had a different trajectory, beginning as weapons manufacturers and subsequently moving into motorcycle and car production.

But with BMW it was a different story, and one that almost immediately plunged the fledgling company into an area of operation where extreme performance and technical sophistication were not just luxuries but essential for survival. For BMW, initiation took place in the heat of aerial combat on the western front in the Great War.

The need for a successful aero engine to combine extraordinary maximum power with minimum weight and faultless reliability when pushed to extremes prompted BMW's founding designers to seek sophisticated engineering solutions, exotic materials, and novel manufacturing techniques. This approach was innovatory in placing quality and performance above cost and transitory convenience. And it was this approach of complete engineering integrity that would be the key to BMW's ascending reputation as the century unfolded, and that would underpin landmark designs such as the R32 motorcycle, the prewar six-cylinder 328 sports car, and even the Neue Klasse sedans of 1961—not to mention the host of immaculately engineered products that by 2005 had powered BMW to its present standing as the global leader in the market for premium vehicles.

The Engine Builder, the Plane Maker, and the Deal Maker

The three separate elements that, in 1917, combined to form Bayerische Motoren-Werke, or BMW, could trace their origins to well before World War I. Karl Friedrich Rapp, an entrepreneurial engineer, set up shop as an aero engine builder in Munich, next to aircraft maker Gustav Otto, son of the man who gave his name to the four-stroke engine. Rapp began supplying Otto with four-cylinder water-cooled engines of his own design, for both aircraft and marine use, but before long came an urgent request to manufacture V-12 aero engines designed by Ferdinand Porsche at Austro-Daimler under a lucrative wartime subcontract. On March 7, 1916—the date regarded as the official beginning of the enterprise that became BMW—Otto's company was reconstituted as Bayerische Flugzeugwerke AG, while the following year Rapp Motorenwerke became known as

9

Where it all began: Gustav Otto's aircraft works at Oberwiesenfeld airfield in Munich. BMW's modern headquarters is still on the same site.

›

Bayerische Motoren-Werke GmbH; the two were combined in 1922 to form BMW AG.

Two men brokered the aero engine deal: Franz Josef Popp, a talented engineer who, even at the age of thirty-two, was already very well connected in Viennese military and financial circles; and Camillo Castiglioni, a colorful Vienna-based financier. Popp was to play an important part in BMW's development until he was forced out by the Nazi Ministry of Aviation in 1942 in a disagreement over aero engine production quotas. Castiglioni's wheeler-dealings saw him successfully finance BMW's early expansion—as well as help strike the key deal to build the Austin 7 car—before his exit in 1929.

With Rapp's early aero engines having been less than successful, he was persuaded by Popp to hire Max Friz, with whom Rapp had worked at Daimler in Stuttgart. Friz, too, was a clear engineering talent, having worked on both Grand Prix cars and aero engines. His first engine for BMW was to be a spectacular success and was the first product to wear the blue-and-white roundel badge.

The six-cylinder Type IIIa, which Friz had already begun to design while at Daimler, was best known as the engine that powered the Fokker D.VII fighter, making it one of the very best combat aircraft of the Great War. Most notably, the IIIa performed well at altitude thanks to its special carburetor, giving

Three key men: aspiring engineer Karl Friedrich Rapp (left); master manager Franz Josef Popp (center), who would steer BMW until 1942; and engine genius Max Friz (right), creator of high-altitude aero engines and the famous boxer motorcycle.

›

10

a first hint of its later potential as a peacetime record breaker in a succession of ever more advanced derivatives.

The IIIa already displayed the eagerness to embrace advanced design and new materials that would characterize BMW's product design into the twenty-first century. The only problem was finding the materials and the space to build the engines: the contract to build the Austro-Daimler V-12 was occupying all the facilities on the Oberwiesenfeld airfield site where BMW still has its headquarters today, and a new assembly hall would not be ready for several months. Thus the air ministry, convinced by the six-cylinder engine's performance, made the decision to allocate V-12 production to Opel in order to enable BMW to produce the IIIa.

Following the armistice in 1918, however, BMW was forced to completely restructure its business away from anything connected with the military; the production of aero engines was expressly forbidden. Thus, following the departures of both Rapp and Otto, the firm began contract production of railway brakes and rejigged the IIIa as a four-cylinder engine

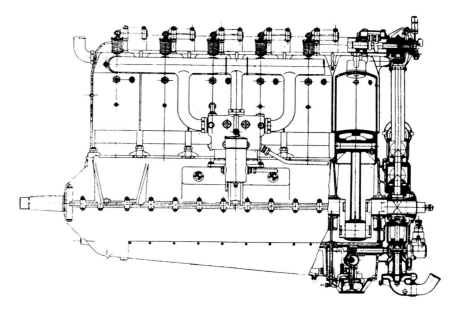

∧ **The straight-six BMW IIIa and IV engines were developed by Max Friz for high-altitude performance, enabling pilot Franz Zeno Diemer to reach a record 9,760m in 1919.** ∨

BMW made frequent use of altitude record breaking in its advertising.

for industrial and agricultural applications. Yet all along, on the quiet, Friz had been working on the IIIa and its Type IV derivative to boost their high-altitude performance still further. He reasoned that while the war reparations forbade the *manufacture* of aero engines, the rules said nothing about developing or improving those existing engines.

Friz's long sessions in the workshop proved to be time well spent, for in June 1919 test pilot Franz Zeno Diemer began his series of altitude-record breaking flights using BMW power (see Chapter 2). Yet while these well publicized achievements did much to enhance BMW's standing in the German aircraft industry, and indeed with the general public, the uncomfortable reality remained that with the cancellation of all military contracts there was huge excess manufacturing capacity across Germany and Europe. Postwar inflation was threatening to turn into hyperinflation, and BMW was relying on tenuous, small manufacturing subcontracts and the future resumption of aero engine production to take it forward. That, at least, was the outward position, but Popp—who by now was in full charge as general director—already had a longer-term plan that aimed to give BMW three main lines of business for genuine stability. His recipe was to continue the production of the profitable industrial engine pending the resumption of aero engine manufacture, and to supplement this business with the development of a motorcycle, which would in turn lead to a move into the car market. The company had already taken a small step in the motorcycle direction when Friz designed a much smaller portable industrial engine. For simplicity, balance, and cooling, Friz chose the layout of an air-cooled flat twin of 500cc—and this was the genesis of the famous "boxer" configuration that would bring BMW motorcycles racetrack glory, commercial success, and an unmatched global identity extending well into the following century.

Yet in 1919, with the flat twin engine predating any motorcycle construction under the BMW name, it fell to other firms to be the first to use Friz's motor in a two-wheeled application. First Victoria motorcycles, then Helios—which was connected with Bayerische Flugzeugwerke before it merged into BMW AG in 1922—installed the flat twin. Friz, however, knew instinctively that the Helios's layout, with the flat twin stretched lengthways in the frame, was wrong: the rear cylinder would be starved of cooling air and would run too hot. His

quick solution was to improve the substandard cycle parts of the Helios machine, sell the existing stock, and move on to design a proper BMW bike that reflected his long-held vision of advanced motorcycle design.

It is difficult from today's perspective to imagine the magnitude of the impact the BMW R32 must have had on the crowds who thronged to see it displayed at the 1923 Paris show. As a bold step into the future, an abrupt break with old-fashioned habits that owed more to nineteenth-century bicycle craft than modern engineering, it must have been as much of an eye-opening sensation as the Citroën DS was in 1955, the Austin Mini in 1959, or, for that matter, BMW's own thrillingly futuristic i8 plug-in hybrid sports car in 2013. With the R32 the motorcycle had suddenly come of age, maturing into a coherent and fully integrated modern product with a design language all its own. Overnight, most other machines, composed as they were from a hodgepodge of proprietary components bought in from traditional suppliers, looked accidental and improvised. It was a tantalizing glimpse of a new era.

Max Friz's breakthrough was to mount the flat twin's cylinders across the frame—as is now familiar from countless generations of BMW R-series motorcycles—and to package the engine, the gearbox, and all the most important ancillaries within a smooth and attractive unit-construction aluminum casting. Most importantly of all, with the engine's crankshaft now inline with the machine's axis, it became natural to drive the rear wheel with a simple shaft, at a stroke doing away with the oily and troublesome mess of an exposed chain or belt to the wheel.

Friz's elegant engine module was housed in a chassis of equal stylishness. Forming a smooth triangle around the engine and with its apex at the rear wheel, the smart black double-cradle frame wore its white pinstriping and its high-mounted BMW badge with pride. The design scored on many fronts—the R32 was neat to look at, its mechanical layout allowed a low center of gravity and a low seat height for smaller riders, and it made for better reliability. And with enclosed valve gear and wet-sump rather than total-loss lubrication, it was much easier to maintain and keep clean.

Though priced at a premium compared to most other machines in the 500cc class, the R32 was an immediate success, capturing the popular imagination. BMW was quick to follow up with a steady stream of improvements and fresh derivatives, including sportier models with overhead-valve versions of the boxer engine. There was even a single-cylinder version—still retaining the classy shaft drive—and before long BMW bikes were winning Grands Prix and, in the hands of works rider and development engineer Ernst Henne, vying

for the title of fastest motorcycle on earth. Most crucially, however, BMW production bikes achieved their considerable variety in the marketplace by the astute use of different engine variations. The core engine, though steadily improved, remained common, and the same tubular frame was shared by all models. But—again as a foreshadowing of what would characterize BMW's later bikes and cars—the BMW look remained the same and its appeal was, if anything, enhanced.

In parallel with their highly bankable work with motorcycles, Friz and his design colleagues had found time to map out a fresh generation of aero engines (see Chapter 2) in time for the resumption of production in 1925. They had also turned their thoughts to the eventual long-term goal of moving into the car business.

The company had for some while been testing prototypes of an advanced small car designed and built by the noted aerodynamicist Wunibald Kamm. The board, however, seemed incapable of making any decision, and in July 1928 the idea was shelved. It was suggested by some,

⌄ Sir Herbert Austin in the tiny Austin 7, which would form the basis of the Dixi, BMW's first car.

perhaps maliciously, that the decision was influenced by board members who also sat on the board of Daimler-Benz. The suspicion was that the famous names of Daimler and Benz, who had merged in 1926, privately feared the likely impact on their business of a BMW passenger car.

The eventual decision was that BMW should indeed build a car, but not one that would compete with Daimler-Benz. Instead, it would be small and of good quality, and preference was to be given to taking over the production of an existing model.

Thus, in October 1928, BMW took over Dixi-Werke, based in Eisenach, some 400km north of Munich. Had BMW known that the Wall Street crash and the subsequent Great Depression were just twelve months away, the directors might have thought differently, but Dixi offered just what Popp and his fellow strategists wanted. Dixi had secured a licensing agreement with Austin of England to build the tiny Seven four-seater, and the lightweight model, renamed Dixi DA1 in its German incarnation, had been selling well since its launch in December of the previous year.

∧ **BMW works rider and development engineer Ernst Henne with one of his speed-record machines in the 1920s.**

The 1930 BMW 3/15 DA3, improved significantly from its Austin origins. ∨

15

CHAPTER ONE

∧ **The 3/20 replaced the 3/15 in 1932 and featured an overhead-valve engine and much-revised chassis.**

‹ **An early Dixi, showing how narrow the Austin-based design was—even for its era.**

∨ **The 3/20 was notably smarter and better built than the lightweight Austin 7 and Dixi models.**

It was the perfect move for BMW, allowing it to apply its undoubted management and engineering expertise to a product that, though innovative and appealing in concept, was rough at the edges and built down to a very low price. By early 1929 BMW had made improvements to the DA1's body, its windows, and, crucially, its braking system. By the summer of that same year, the model was wearing BMW badges and had become much more substantial in its appearance and construction.

Friz, however, was becoming frustrated with the clear limitations imposed by the very basic Austin hardware. In the course of his work on the DA1, he had managed to add independent front suspension and an overhead-valve update of the still tiny engine. At the same time it was redesignated 3/15 to signal the important engineering updates. But even with the Depression beginning to take hold, there was a sense within BMW that it might achieve better results with a more modern car of its own design. In 1931 Friz proposed and tested a front-wheel-drive design using a twin-cylinder two-stroke engine, but its performance was judged disappointing, and he resigned to take up a position at Daimler-Benz. Popp urged his new team to come up with a fresh and entirely Austin-free design, and by mid-1932 the AM-1 (the initials standing for Automobile München) was ready. Known commercially as the 3/20, it had a new backbone chassis, independent suspension all round, and an all-steel body initially manufactured by Daimler-Benz. The 3/20 was larger than the Austin-derived 3/15 that had kept BMW sales going, but though its space and refinement gained praise, contemporary commentators were less sure about its steering and handling—something that would later be addressed and that would become a powerful part of BMW brand identity.

Even though the 3/20 was a qualified success in those difficult times, the engineering team under Alfred Böning continued to evolve the design, giving the gearbox a fourth speed. More prophetically, however, Böning had been encouraged by the sporting successes of the 3/15 and was struggling to improve the performance of the still Austin-based engine. Rudolf Schleicher, who had become head of engine testing in 1931 after a spell at luxury carmaker Horch, suggested that rather than develop an all-new four-cylinder unit, BMW should simply add two more cylinders to take the existing engine to 1.2 liters and 30 hp. It was thus that, at the Berlin Motor Show in February 1933, the six-cylinder BMW 303 made its public debut.

Though still an upright-looking sedan with a sober air, the long-hooded 303 was BMW's largest car so far, as well as the first to house a straight six-cylinder engine and sport the double-kidney grille topped with the BMW roundel. And with the arrival of these vital ingredients, the stage was set for the BMW story to really begin.

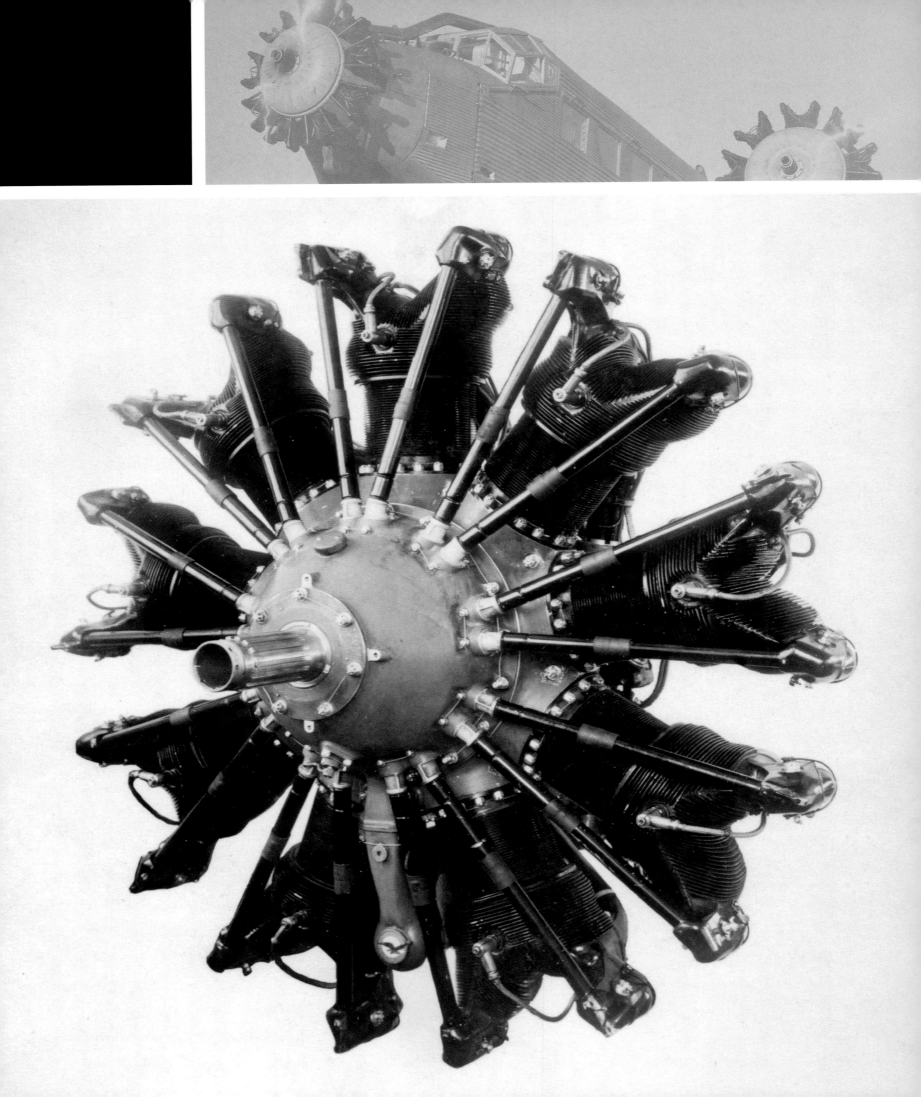

2

The Power to Fly

BMW IIIa, 1917: BMW Prepares for Takeoff

BMW's first aero engine (pictured page 11) was remarkable not only because it was arguably the highest-performing aircraft engine of the First World War but also because it gave rise to the founding of the BMW company itself.

In 1913 Rapp Motorenwerke was established in Munich by chief designer Karl Friedrich Rapp; however, his engines proved unremarkable, and in 1917 designer Max Friz was brought in to draw up a new engine. Friz's design used the same six-cylinder inline configuration as the Rapp III, but the result proved infinitely superior. Rapp left the company, which was subsequently renamed Bayerische Motoren-Werke GmbH; the Friz design was designated BMW IIIa, and volume production started early in 1918.

The smooth-running, 201-hp, 19.1-liter, water-cooled inline six-cylinder engine was light, with an aluminum crankcase and pistons. One-piece cylinder heads and cylinders did away with the need for head gaskets—the valves were opened by a single overhead camshaft driven from the crankshaft by a vertical shaft.

The breakthrough IIIa was a high-altitude design successfully running a relatively high compression ratio of 6.4 to 1 and fueled by a "high-altitude carburetor." This comprised three mixing chambers, three air and fuel jets, and five throttle butterflies. Two levers in the cockpit allowed the pilot to set the engine for low and high altitude. High altitudes were crucial to gaining an advantage in air combat, and when the IIIa was fitted to a Fokker D V.II, it could outclimb all the Allied opposition.

BMW IV, 1917: BMW Flies High and Ramps Up the Volume

The BMW IIIa was easily the best and highest-performing aero engine of World War I, but, luckily for the Allies, it arrived too late. In September 1919 BMW's second aero engine set a world record for a passenger aircraft when Franz Zeno Diemer flew eight people to an altitude of 6,750m in a Ju F.13 monoplane powered by this engine.

Prior to that, in June 1919, Diemer broke an "absolute" altitude record by flying a DFW F37/III single-seater biplane to an altitude of 9,760m. The aircraft was not powered by a BMW IIIa but by an evolution of it, the BMW IV (pictured previous page). Both engines were water cooled, but the IV had been further optimized compared to its predecessor to produce 241 hp. It took Diemer eighty-seven minutes to climb to the record altitude, a mean rate of climb of 113m per minute. By contrast today's Eurofighter Typhoon, co-developed by European nations including Great Britain, Germany, and France, has a rate of climb of 19,077m per minute.

BMW VI, 1926: First V-12 Is a Round-the-World Success

The end of hostilities in 1918 meant that BMW was prevented from producing aero engines until 1922, and it was 1923 before production resumed on the IIIa and IV. However, the world of aviation was moving apace, and demand quickly grew for higher performance power units. Thus in 1924 Max Friz was instructed to start work on a 46.9-liter, 60-degree V-12 monster producing 644 hp. This would be BMW's first ever V-12 engine, and it was called BMW VI (below). Another advanced, high-performance design, the V-12 used the six-cylinder as a basis, a technique that today would be called a "modular and scalable" approach.

Series production began in 1926, and the engine quickly set a new benchmark. At the start of the 1932, it was used to

power the Schienenzeppelin (rail Zeppelin), an experimental railcar that resembled a Zeppelin airship. On June 21, 1931, the Schienenzeppelin set a world record for a fuel-powered rail vehicle of 230.2km/h. More significantly, it powered a Dornier Wal flown by Wolfgang von Gronau on a transatlantic flight from Sylt in Germany to New York via Iceland, Greenland, and Labrador. The flight established the northern air route over the Atlantic in forty-seven hours.

In 1932, von Gronau flew a second Dornier Wal around the world, powered by a BMW VII that was equipped with a 0.62 ratio-reduction gearing system for the propeller. The V-12 also formed the basis of BMW's first experiments with direct fuel injection in 1933.

BMW 132, 1932: Lufthansa Insists on BMW Radial Engines

In 1929 BMW bought the rights to manufacture Pratt & Whitney Hornet and Wasp radial engines. The arrangement was canceled in 1931, but in 1932 events took a turn for the better. Junkers revealed its new civil airliner, the three-engined Ju 52 (opposite, below), which was originally going to be fitted with Junkers Jumo engines. However, the manufacturer's main customer, Lufthansa, insisted the aeroplane be powered by BMW radial engines. BMW quickly refreshed its license with Pratt & Whitney and developed an improved version of the R-1690 Hornet called the BMW 132. The nine-cylinder engine had a displacement of 27.7 liters. Radial engines are distinct from the rotary engines used during World War I in that the main body of a radial is static and the crankshaft rotates, like an inline engine; in contrast, a rotary engine revolves about a fixed crankshaft.

The civilian engines used on the Ju 52 were fed fuel through carburetors, but BMW produced several versions of the engine and also used it to experiment with direct fuel injection. A large number of derivatives were built, developing 715 to 947 hp, the most powerful being direct fuel injected. As well as the Ju 52, the 132 would go on to power World War II fighting aircraft such as the Ju 86 bombers, Arado Ar 196 floatplanes, Focke-Wulf Fw 200 patrol bombers, and several others. Prior to hostilities breaking out again in 1939, the first nonstop flight from Berlin to New York had been made by an Fw 200 Condor S-1 powered by four BMW 132 engines.

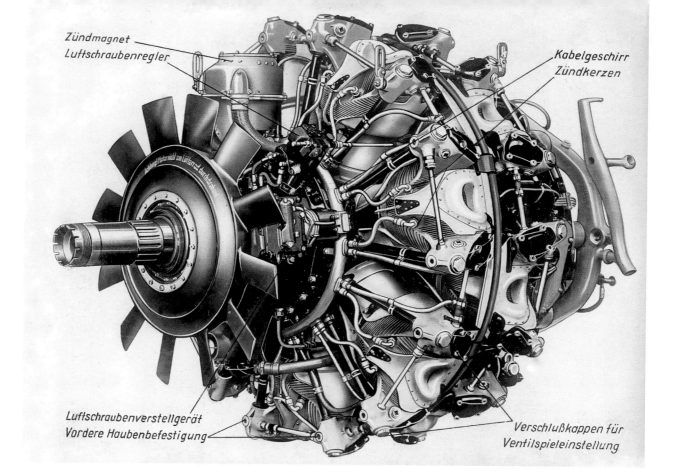

Zündmagnet
Luftschraubenregler

Kabelgeschirr
Zündkerzen

Luftschraubenverstellgerät
Vordere Haubenbefestigung

Verschlußkappen für
Ventilspieleinstellung

BMW 801, 1939: The Feared and Revered Two-Row Radial

One of the most respected and feared fighting machines of World War II, the formidable Focke-Wulf Fw 190, was powered by one of BMW's finest ever aero engines: the air-cooled, twin-row, fourteen-cylinder 801 radial (above). More than twenty-eight thousand of these engines were built during the course of the war, and all delivered phenomenal performance.

BMW acquired a competitor, Bramo, in 1939; the 801 was the result of merging the BMW 139 with the Bramo 329, both of which had been funded by the German Ministry of Aviation in 1935. Surprisingly, the 801 was equipped with just two valves per cylinder, while the inline engines of the time were already using four valves per cylinder to improve engine breathing and performance. The 801, however, did feature sodium-cooled exhaust valves and direct fuel injection.

The initial design, which first ran in 1939, was for a pair of engines—the 801A and 801B—to be used in twin-engined aircraft. The engines were equipped with gearboxes that rotated the airscrews in opposite directions, canceling out the torque effect. The 801C, with improved cooling, was the first power unit fitted to the Fw 190.

The 801 was continually developed as the war progressed, with the performance of the British Spitfire, powered by the Rolls-Royce Merlin, and that of the Fw 190

21

leapfrogging one another with each successive advance. BMW improved its original supercharger to increase performance at higher altitudes, and the power outputs of operational 801s eventually rose to a mighty 2,400 hp. Ultimately, the 801 would power a range of aircraft in addition to the Fw 190, including the much-feared twin-engine Ju 88 fighter-bomber.

BMW 803, 1942–1944:
A Twenty-Eight-Cylinder Monster

The BMW 803 (above) was an ambitious project consisting of two 801 radial engines grafted together and water cooled to avoid problems of overheating. The engine was intended to power larger versions of the Focke-Wulf Ta 400 long-range bomber and other similar multi-engine aircraft, as well as single-engine fighters.

In order to avoid reliability problems and other difficulties associated with engineering a single long crankshaft, the 803 became fiendishly complicated. As the 801 was a twin-row fourteen-cylinder engine, the 803 had twenty-eight cylinders arranged in four rows, but each pair of rows retained separate crankshafts inline with one another. The front unit drove the forward airscrew directly, but the rear unit drove the rear airscrew through a series of shafts and a gearing system. As a result, a large gearbox was attached to the front of the engine, a hollow shaft driving the rearmost airscrew, with the crankshaft from the forward fourteen cylinders passing through it.

A further challenge in having two separate crankshafts was how best to drive ancillary components. The most important of these was the supercharger, which on an engine of this size consumed hundreds of horsepower. The decision

was made to drive it from the rear engine, reducing the power available to drive the rear airscrew, but the engine still delivered a substantial output of 3,847 hp. This made it the most powerful German piston engine to date, but it was so heavy that the power-to-weight ratio was poor compared to alternatives. The 803 never entered production, and with the resources of the Nazi war machine steadily dwindling, neither did the aircraft for which it was intended.

BMW 003, 1944: BMW's First Jet Engine
Ends the Piston Era

Both Sir Frank Whittle and Hans von Ohain are credited with the invention of the jet engine, Whittle securing a patent for his centrifugal design in 1930 and Ohain for his axial principle in 1936. However, it was the German engine that flew first, in 1939, with the Whittle unit not making its first flight until 1941. Eventually the axial design would be adopted by the world's jet engine manufacturers in postwar years.

The BMW 003 engine (opposite, top) was equipped with a seven-stage compressor and sixteen burners set into an annular combustion chamber. First tests of the 003 took place in 1940, with the engine attached to the underside of a twin-engine Messerschmitt Bf 110. Disappointingly, it only produced 150kg of thrust—half the expected amount. Development was slow, partly because of the scarcity of the very high-grade metals needed to build the internal components of a jet engine, such as nickel, cobalt, and molybdenum.

A second test took place using a modified twin-engine Messerschmitt Me 262, also equipped with a conventional BMW 801 "safety" engine attached to the nose. The test

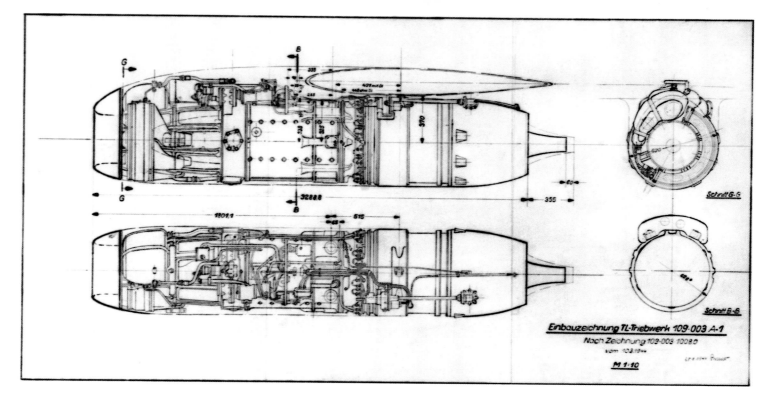

was a disaster, and the aircraft only just made it back to the airfield using the piston engine. Development continued, and, following a more successful test in 1943, the 003 was readied for production in 1944. The engine proved unreliable, however, and the production twin-engine Me 262 was instead powered by two Junkers Jumo jet engines. The 003 was fitted to the Heinkel He 162 and the four-engine Arado Ar 234C, but only around five hundred examples of the jet were built.

BMW J79-GE-11A and BR700 Series Jet Engines

BMW was involved in two jet engine projects after the war. The first was as part of a consortium that delivered the General Electric J79-GE-11A engine used to power the Lockheed F-104G Starfighter. The second was a short-lived association with Rolls-Royce to build the BR700 regional jet engine (right).

The Starfighter engine developed 8,119kg of thrust with afterburner, enough to blast the controversial and tricky-to-fly fighter to 2,235km/h, or Mach 2.2. The J79, an axial-flow turbojet engine like the original BMW 003, remained in production for thirty years. As well as being built under license in several countries, it was fitted to a number of different aircraft, and seventeen thousand were made in the United States alone. The axial-flow design was eventually superseded by quieter and more efficient turbofan engines in large passenger aircraft.

BMW returned to aero engine manufacturing in 1990 and founded a joint venture with Rolls-Royce plc, called BMW Rolls-Royce AeroEngines GmbH, to build the BR700-Series civil aircraft engine. The joint venture was short lived, however, and Rolls-Royce took over the company completely in 2000.

Three main versions were developed: the BR710, BR715, and BR725. The first two are both twin-shaft turbofan engines. The 710 is equipped with a 1,220mm-diameter single-stage fan driven by a two-stage turbine, while the BR715 has a 1,321mm turbofan driven by a three-stage turbine. The BR725 was designed for the Gulfstream G650 and has a three-stage, axial-flow, low-pressure turbine. It has a 1,270mm fan and develops 7,711 kg of thrust.

23

3

The Power of Six

The year 1933 was a turning point for German industry and, though much more darkly, for the German nation. For that matter, it was a turning point for the whole world: Adolf Hitler's seizure of power at the beginning of the year would have catastrophic consequences for humanity as the decade unfolded. It is also an uncomfortable truth that in the early days of Nazi rule, the country's business leaders gave a guarded welcome—if only through gritted teeth—to many of the measures announced by the outspoken new *Reichskanzler* (chancellor).

Mass mobility and popular car ownership had always been prominent among Hitler's pet projects. Within two weeks of installing himself in office, he opened the Berlin Motor Show, where the shiny new creations perfectly fitted his vision of a glorious, technologically driven future with German industry secure in the driving seat. Days later, he summoned the leaders of the country's five surviving carmakers—times had indeed been hard—to announce his four-point plan for national car ownership. Among those leaders was Franz Josef Popp of BMW.

What those industry leaders heard must have been music to their ears, even if the tone was strident and intimidating. With car sales stagnant after several years of national austerity, economic decision making in paralysis, and the industrial outlook uncertain, the lifting of

∧ Germany's Autobahn offered the first opportunity for sustained high-speed driving.

⟨ Bred to win: the high-performance 328 roadster broke new ground in style, sophistication, and speed.

▲ **The 303, presented in 1933, was BMW's first six-cylinder car—though the initial engine's displacement was just 1,173cc.** ⌄

taxation on cars promised a quick stimulus to sales. More importantly, the National Socialists' plan provided long-term security and certainty—qualities always favored by investors—at a time when, for the past decade under the Weimer Republic, dithering and stopgap measures had been the order of the day.

Indeed, this plan would lay the foundations for Germany to become the pre-eminent automobile engineering nation in the world—a status that survived the horrors of the Third Reich and that arguably survives intact today. Among its provisions were the building of a national *Autobahn* network to vie with fellow dictator Benito Mussolini's expanding *Autostrada* system in Italy, and the construction of motorsports facilities, among which would be the dramatic Nürburgring, carved out of the Eifel mountains. Feeding directly into this grand infrastructure plan would be a rationalization of the components industry and a fresh generation of powerful cars, newly liberated by the removal of the horsepower-based tax and honed in the much higher-speed environment of the new motorways. This would encourage larger engines, boost pride and prestige, and, in turn, increase the prospects for exports—especially to the power-hungry United States.

Whether by coincidence or through prescience, this was just the time when BMW, struggling to eke a few more horsepower out of the tiny Austin-derived four-cylinder it had been relying on for too many years, had begun developing a six-cylinder derivative to power a new wave of medium (as opposed to small) passenger cars. The 303, presented as a prototype at February 1933's Berlin Motor Show, not only looked more elegant with its clearer, slightly sweptback double-kidney grille, graceful long hood, and flowing wings—it was also the first of the company's cars to house a six-cylinder engine. This asset would become synonymous with the BMW brand for decades to come and would form the true foundation for every generation of BMW that followed.

That might seem an unlikely achievement for a simple engine displacing just 1,173cc and giving a meager 30 hp. But engineers Alfred Böning and Fritz Fiedler (the latter having followed Rudolf Schleicher from Horch in Zwickau) had planned ahead and had placed the engine's bore centers further apart to allow room for later expansion. And with the big stretch in wheelbase giving the 303 a much longer hood as well as a roomier passenger compartment, the soon-to-be-familiar cab-rearward proportions had begun to emerge. In response to earlier criticism of BMWs' steering responses, the 303 adopted more sophisticated suspension geometry as well as hydraulic brakes and rack-and-pinion steering; though, again, it had an initially rocky ride because of mismatched front and rear spring rates. Even so, a recognizably modern car was clearly taking shape, and development proceeded rapidly.

As further body styles emerged, the skilled hand of chief designer Peter Schimanowski began to shine through. The first real hint of the delights to come appeared with the 315/1, the two-seater sports-roadster version of the staple 315 that replaced the 303 in 1934, using an enlarged 1.5-liter version of the small six. The 303 was strongly priced and had sold well, more than doubling BMW's annual sales between 1932 and 1934, but it lacked true distinction. The 315 in its everyday sedan and convertible versions saw a marked step up in both style and engineering refinement. Yet, less than five years after BMW's decision to enter the car market, it was that two-seater, the 1935 315/1, that really put the company on the map for elegance, sporting performance, and overall desirability.

27

▲ The stylish Schimanowski-designed 315/1 and 319/1 put BMW on the map for elegance and desirability.

Shimanowski's rakish roadster design ticked all the right boxes, and many of its qualities still resonate today. The rearward-slanted twin-kidney grilles and long, side-louvered hood suggested streamlining and speed, as did the sloped windshield and the gracefully tapering tail. The cutdown doors gave a good view into and out of the cockpit, if only to ensure the driver and passenger were suitably admired out on the street. And the steel-disc wheels (in place of old-fashioned wire-spoked hubs) combined with full fairings over the rear wheels to present a clean and modern look. An elegant two-tone color scheme completed the picture.

In direct contrast to its predecessors, the 315/1 was not "all show and no go." It was lighter than the sedans, and higher compression and triple carburetors boosted the engine to 40 hp. BMW managers were quick to realize its potential for competition and, with a view to the 1.9-liter, 55-hp version soon to be on the stocks, the company began backing privateer drivers in events such as the International Alpine Trial.

In parallel with the six-cylinder 315 and 319, BMW was equally successful with a four-cylinder version, the 309—so successful, in fact, that the company enjoyed a 6 percent share of the German market in 1934, running in fourth place overall, ahead of Daimler-Benz. This success, allied with the growing demand for motorcycles and Hitler's massive rearmament program swelling the order books for aero engines, was the perfect tonic for BMW's balance sheet. Nevertheless, it posed something of a problem for the car division: all the design and manufacturing facilities were concentrated at the Eisenach plant the company had inherited from Dixi. The factory was running at full capacity, and the only way of expanding output would have to be to invest in a new plant. It was at this point that Popp made a key strategic decision that would reinforce BMW's standing as a brand for the rest of the century: rather than building a new plant to make more cars, BMW would move its cars upmarket in terms of size, power, prestige, and price, thus improving its profit per vehicle and ensuring that supply always trailed slightly behind demand.

It was a policy shift that played directly to BMW's strengths, especially those of Fritz Fiedler, in chassis engineering, and Alex von Falkenhausen, a young nobleman who had studied automobile and aircraft engineering under

Setting a style: models such as this 1937 327 coupe established BMW cars as elegant and desirable.

Fritz Fiedler (far left) and Alfred Böning (center), architects of a generation of BMW six-cylinder engines, with racer engineer Alex von Falkenhausen (right). All three would have a profound influence on the company.

‹

Willy Messerschmitt. Fiedler and von Falkenhausen would have a profound influence on all spheres of BMW engineering right into the 1960s. This top team, which also included Böning and Ernst Loof (the latter of whom, like von Falkenhausen, had been a works BMW motorcycle racer) went on to create a bewildering variety of models in the immediate prewar period. These included such all-time classics as the 326, in its many bodywork variations, and the glamorous 328 roadster, one of the most instantly successful competition cars the industry has ever known.

The BMW engineers had at their disposal what their modern counterparts would call a modular toolkit: the six-cylinder engine in an assortment of sizes and states of tune; a newly developed chassis in various lengths and of far greater rigidity than the previous tubular affairs; new torsion-bar rear suspension; revised independent front suspension with the transverse leaf spring now at the bottom; and, finally, the great variety of body styles possible through the creations of outside *carrossiers* and BMW's in-house design team. The

∧ The new generation of larger cars began with the 326 and had the benefit of a much more rigid chassis, with hydraulic brakes and rack-and-pinion steering. ∨

30

most commercially successful of these new models was the 326, launched in early 1936. Its engine, now almost 2 liters in capacity, gave a smooth and gentle 50 hp; the four-speed gearbox now featured synchromesh and freewheel; and hydraulic brakes and rack-and-pinion steering were standard. Most vitally, however, Schimanowski had come up with a compelling new style, the now-characteristic double-kidney grille smoothed back into the long hood and the flowing wings. The low and lithe look was even more skillfully exploited with the later 327, essentially a coupe version of the 326, where the low roofline perfectly complemented the long drawn-out hood, the now faired-in headlights and the wide, grippy stance on the road. It was the first BMW fixed-head car with a genuine intrinsic visual allure, joining the two-seater roadster in distinguishing BMW as a prewar style setter.

By now BMW's four-cylinder cars had been dropped and the upward push continued. Fiedler and his team began work on an even larger sedan, the 335, to be powered by a 3.5-liter evolution of the familiar six, now using a spur-gear drive to the high-mounted camshaft instead of the usual duplex chain. Though first shown in prototype form in 1938, the 335 only entered production the following year, and relatively few were built before the enforced stop on car production in 1941.

And yet nothing, not even the elegant 327 or the knowledge that the luxurious 335 was in the pipeline, could have prepared the automotive world for the sensation that the 328 sports roadster would instantly become. Capitalizing on the still racy reputation of the 315/1 and 319/1 two-seaters, Popp had drawn up plans for a more modern replacement. Work began on an ambitious upgrade of the 2-liter engine, with an aluminum-alloy cylinder head housing thermodynamically efficient hemispherical combustion chambers—which in turn required splayed valves and intricate arrangements of pushrods and rocker arms and shafts to actuate them from the high-mounted camshaft. With triple carburetors, a high 7.5:1 compression ratio, and a safe rev limit beyond 5,000 rpm, the engine worked brilliantly well even in its standard production form, giving an easy 80 hp. Competition developments would soon see output pushed well into three figures.

Matching the 328's engineering specification, which also included a full-length undertray to smooth out airflow

under the chassis, was a two-seat roadster body that was quite simply breathtaking. Smooth, aerodynamic, and elegantly proportioned with its blended-in double-kidney grille, large headlights—also neatly faired in—and its stylishly chamfered boat-tail, it looked every inch a racer for the road. Visual clues were provided in the shape of two Le Mans–style leather hood straps and the cutdown cockpit doors to allow the driver elbowroom to work the steering wheel in the heat of competition.

The 328 got off to a dream start, with the prototype winning its first race—the prestigious Eifelrennen at the Nürburgring in June 1936—driven by Ernst Henne, who beat a whole field that included supercharged racing cars. When road-going 328s began to reach customers later that summer, their maximum speed of over 150km/h immediately made them off-the-shelf race winners, too, and before long the 328 began notching up wins in close succession. The model dominated road and circuit racing as well as the then high-profile sport of hill-climbing in 1937. It took class wins in the 1938 Mille Miglia and German Grand Prix, and it also proved the fastest sports car in that year's Avusrennen.

The following year saw a works team enter the 24 Hours of Le Mans, with a special-bodied streamliner coupe winning the 2-liter class and coming in fourth overall and two standard-bodied

The 328 sports roadster was a sensation when it debuted in 1936, with stunning looks and race-winning performance from its 2-liter engine uprated to 80 hp. ❯

Ingenious and intricate valvegear design was key to the 328's remarkable performance; triple carburetors helped, too.

roadsters finishing seventh and ninth overall. The 328's winning streak even continued into wartime, with Fritz Hushke von Hanstein, later competitions director at Porsche, scoring an outright win in the 1940 Mille Miglia run over an amended course.

All this was the best possible advertisement for a car that was always heavily in demand, and whose instant-legend status was out of all proportion to the tiny numbers actually produced (between 410 and 462—sources disagree). Putting aside the general hysteria surrounding it, however, the 328 was an important engineering milestone for both BMW and the whole industry. It was convincing proof that sports cars did not have to be large, heavy, and massively powerful to be successful. It showed, too, that a car—any car—with a stiff and well-engineered chassis could both be comfortable and civilized to drive under everyday conditions and perform astonishing feats on the racetrack.

Thus, with its reputation sky high, its civilian car and motorcycle business booming, and its aero engines in round-the-clock production for the Luftwaffe, BMW was plunged into the unknown of total war. Only one thing was truly certain: the company's operations would no longer be directed by its own skilled managers but by politically enforced diktat from Berlin.

⌃ Private and factory-entered 328s swept the board in all forms of motorsport from the 1937 season on, with wins at Le Mans, Ulster, the Mille Miglia, and other circuits; final cars had special coupe bodies by Touring.

4

The Boxer Legend

The 500cc horizontally opposed flat twin that was the cornerstone of BMW's motorcycling success started life as an industrial engine. It was originally fitted, to little effect, to the short-lived Helios motorcycle built by sister company Bayerische Flugzeugwerke. The masterstroke of BMW's trio of talented motorcycle engineers was to turn the engine across the frame so that both cylinders received cooling air, and so that an efficient straight driveline could be taken to the rear wheel via a shaft drive—the second stroke of genius.

Thus was born the first BMW "boxer" motorcycle, which sent shockwaves through the industry and established the template for almost a century of design. But BMW did more than just bring the flat twin and shaft drive to the motorcycle sector. Its smoothly styled models brought an integrated approach to motorcycle design, getting rid of extraneous clutter, eliminating oil leaks, and helping dispel the grubby, oily-rag image suffered by even the more affluent two-wheeled customers of the era. Most importantly, the clever engineering and the little luxuries were not confined to the expensive, large-capacity bikes but were shared across the lineup—something that still holds true today.

BMW R32, 1923

To describe the R32 (right) as a landmark in motorcycle design would be something of an understatement. The public reaction on its unveiling in 1923 was one of utter amazement, as if a

vision of the future had suddenly landed. With its smoothly integrated frame and tank carefully wrapped around the unit-construction flat-twin engine and its neat shaft drive to the cast-aluminum rear hub, it was the first motorcycle to be designed as a coherent whole. At a stroke it made everything else look bitty, amateurish, and dated, and it represented a decisive step away from the bicycle-era engineering that had for too long been the norm. Though modest in its power output at 8.5 hp, the R32 was smooth, reliable, and clean, earning it an instant following. Profoundly influential on both BMW and the rest of the industry, it signaled the systematic and professional approach to engineering that very soon became the core ethos of the company.

BMW R39, 1925

Every bit as smart and stylish as the R32 that inspired it, the R39 (not pictured) was BMW's first single-cylinder bike. With its 250cc engine mounted upright in the frame and giving 6.5 hp, it had the distinction of being the first motorcycle to boast a fully enclosed overhead-valve cylinder head. It also had the novelty of a one-piece light-alloy casting for the cylinder barrel and crankcase, with a pressed-in cylinder liner. Crucially, this smaller bike retained the smooth, maintenance-free shaft drive of its bigger brother, emphasizing its engineering integrity.

BMW R37, 1925

The R37 (above) was BMW's first venture into sports motorcycle territory. Based directly on the R32, it retained the same 494cc capacity but introduced the major technical innovation of overhead-valve cylinder heads. This allowed a rise in compression ratio to 6.2:1 and a near doubling of power, to 16 hp at the then-giddy speed of 4,000 rpm. With a total weight of just 125kg this produced sparkling

performance—enough, in fact, to win the German Grand Prix in its first year and to take home gold in the 1926 International Six Days Trial in England, with young engineer Rudolf Schleicher aboard.

BMW R62 and R63, 1928

Essentially the same motorcycle save for the valve actuation in the engine, the R62 (opposite, top) and R63 were the first BMW machines in the 750cc class—a category that was becoming more popular because of growing demand for sidecar travel. With the standard side-valve engine and low compression, the R62 gave 18 hp, while the overhead-valve engine in the R63 delivered 24 hp for a maximum speed of almost 120km/h. Up to this point all BMW motorcycles had still featured unsprung rear ends and V-block band brakes on the rear wheel.

BMW R11 and R16, 1929

The R11 tourer and R16 sports bikes (opposite, center) took over from the R62 and R63. Though similar in appearance,

they moved the motorcycling game on—especially on the manufacturing front. Their major innovation was the pressed-steel frame, which gave much better torsional rigidity for sidecar work; like the earlier bikes, they used side- and overhead-valve versions of essentially the same 750cc unit. A further claim to fame was that these models now offered the previously optional lights, speedometer, and horn as standard equipment.

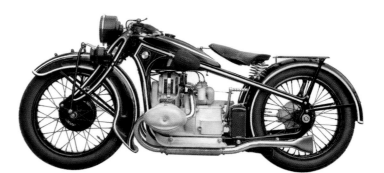

BMW R2, R3, and R4, 1931–1936

These modest single-cylinder machines may have lacked glamour, and they are often overlooked in abridged histories of the motorcycle industry. But for BMW, the 200cc R2 and the later 400cc and 300cc R4 and R3 (lower right) provided solid bread-and-butter business through tough economic times, selling more than thirty thousand units in total. A technical curiosity of this series was the offset mounting of the engine from the chassis centerline. In top gear this gave a highly efficient direct straightline drive from the crankshaft to the rear wheel, something Lexus would spend millions trying to achieve in its first luxury car in the late 1980s.

BMW R12 and R17, 1935

The R12 and R17 mark an important milestone in motorcycle history: they were the first models to feature telescopic front forks, now an almost universal feature in chassis design. Introduced after many years of experimentation with different types of suspension, the hydraulically damped forks provided much longer travel than traditional solutions, improving not only comfort and refinement but also straightline stability and safe handling in corners. Now giving a mighty 33 hp and capable of 140km/h in R17 form, this generation also marked the debut of a four-speed gearbox and an integrated drum rear brake.

International Six Days Trial, 1935

Many thought it highly risky for BMW to enter this prestigious 1,000-plus-mile event with supercharged bikes, especially as the conditions ranged from open roads and speed tests to rough tracks and off-road sections. Added to this were strict rules about carrying spare parts and the riders being responsible for their own repairs. Yet despite being tuned to give more than 50 hp from 500cc, the supercharged BMWs were highly successful in the hands of works riders such as Henne and proved that high performance engines could be as reliable as standard units.

BMW R5, 1935

This high-performance 500 marked yet another engineering milestone. Drawing heavily on BMW's racing experience, it had a new engine, complete with twin chain-driven camshafts and overhead valves as well as an entirely new frame using oval-section tubing welded using an innovative electrical process for much better strength and rigidity. The four-speed gearbox moved over to a foot shift, though an auxiliary hand lever was retained to placate conservative customers, and footboards made way for modern-style footpegs and a toe- (rather than heel-) operated brake. With 24 hp at a speedy 5,500 rpm, the R5 could be seen as one of the first modern middleweight sports bikes. By 1938, rear suspension had also been added.

BMW Speed Records, 1937

Ernst Henne was an enthusiastic competitor in a variety of different forms of motorsport and had begun setting straightline speed records on BMW machines starting in the late 1920s. By the mid-1930s the bikes were specialized 500cc supercharged streamliners capable of more than 250km/h, and by 1937 BMW could claim the title of the fastest motorcycle in the world at almost 280km/h on the A3 Autobahn near Darmstadt. The record would stand until broken by Wilhelm Herz on an NSU special in 1951.

BMW in the Senior Tourist Trophy, 1939

The Senior Tourist Trophy (TT), held on the daunting 60km Mountain Course on the Isle of Man and regarded as the world's most prestigious road race, had long been seen as a stronghold for British bikes and British riders. Eyebrows were raised when Briton John "Jock" West finished sixth in the 1937 event on a supercharged 500cc BMW, but in the 1939 German domination was complete: Georg Meier on a similar machine took the top prize, beating ten-time TT winner Stanley Woods into fourth place. Meier was the first foreign rider on a foreign bike to win the Senior TT event.

BMW R75, 1941

The military R75 combination, developed under the leadership of racer and engineer Alex von Falkenhausen, was a remarkable vehicle. Designed to cross all manner of terrain—from sandy deserts to snow-laden steppes and muddy swamps—the R75 was a full-blown off-road machine with drive to the sidecar wheel as well as the conventional rear wheel. It had an ingenious gearbox with eight forward and two reverse ratios, as well as a differential lock for the sidecar drive, and with a torquey 750cc side-valve engine it was legendary in its load-carrying ability. It was a decisive contributor to the German military's wartime mobility.

BMW R24 and R25, 1948

These were the first bikes built by BMW in the postwar period, when restrictions prohibited engines above 250cc. The stylish R24 (above), developed from the prewar R23 but with a new four-speed gearbox, retained BMW's trademark shaft drive and

was an immediate success. The design was steadily improved as better components and materials became available. By 1950 this model had given way to the R25, featuring plunger rear suspension instead of a rigid frame; this gave a much-needed boost to comfort. Despite its relatively high price, the R25 sold more than 108,000 units before it was replaced by the R26, with swinging-arm rear suspension, in 1956.

BMW R51/2, R67, and R68

The return of the legendary BMW flat twin was greeted with great excitement in 1950, when restrictions on the production of larger motorcycles were finally lifted. The R51/2 (below) was essentially the same as the prewar machine but benefited from improvements to the carburation and camshafts, now driven by spur gears instead of a noisy chain. The plunger frame was retained, but brakes became full width and the damping for the telescopic front forks was improved. The 600cc R67, aimed at sidecar applications, soon followed, and in 1952 the solo R68—with a 35-hp evolution of the flat twin—became the first German production motorcycle to reach 160km/h.

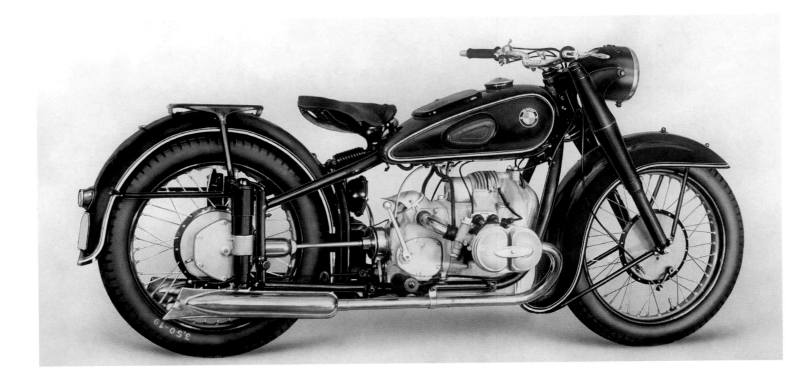

BMW R50 and R69, 1955

Launched just five years after the resumption of flat twin production, the 1955-generation R50 (500cc, above) and R69 (600cc) marked a significant advance in BMW chassis design. Having pioneered telescopic forks in 1935, BMW now moved to a completely new frame with swinging-arm suspension front and rear. Sometimes known as Earles forks, the front leading arms were accurately located in adjustable taper-roller bearings, improving not only comfort and handling precision but also stability and braking: the bike no longer nose-dived when slowed, allowing the rear wheel to share more of the braking effort.

BMW R69S, 1960

The definitive BMW sports bike of the 1960s (above) was a response to the growing success of the racy British parallel twins from Triumph and BSA on the US market; it also benefited from the backing of the new driving force behind BMW, Herbert Quandt. Raising the compression and the rev limit helped produce a respectable 42 hp, and a world-first hydraulic-steering damper tamed the wobbles of the heavy Earles fork frame at speed. However, towards the end of the decade, the R69S bikes—and BMW itself—were beginning to look very conservative as Japanese manufacturers started launching their flashy four-cylinder superbikes.

BMW R27, 1960

The sophisticated R27 single (left) was BMW's answer to the fundamental shift in the motorcycle market that began in the early 1960s. Bikes were becoming a leisure pursuit rather than a transportation necessity, and Japanese makers such as Honda were appearing with fast and well-equipped models, especially in the smaller capacity classes. The R27 sought instead to be smooth and quiet, with a novel elastomer engine and driveline mounting—and, of course, shaft drive. But it lacked performance and sparkle, and as a quality bike it could not compete on price. All BMW singles were dropped in 1966, and buyers would have to wait until 1993 for the next one to appear.

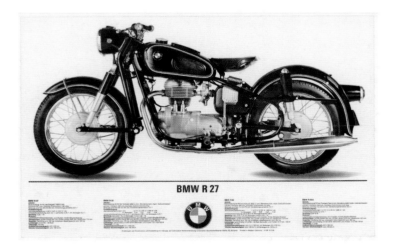

41

BMW /5 Range, 1969

Times became increasingly hard for BMW motorcycles during the 1960s as Japanese bikes seized the growing recreational biking market and ever fewer buyers were willing to pay top prices for what now looked like a slow, expensive, and outdated BMW machine. The company even contemplated an exit from motorcycle production but resolved instead on a much more modern iteration of its trademark flat twin. The /5-series models (above), debuting in 1969, were lighter, more agile, fresher-looking, and more powerful—and, at last, they had an electric start and were available in brighter colors than classic black. These were the first bikes to be made at the Spandau factory in Berlin.

BMW R90S, 1973

Perhaps the biggest ever advance in the public perception of BMW motorcycles came with the debut of the R90S (above) in 1973. As the flagship of the /6 series, it added not only a five-speed gearbox and disc front brakes but also a potent 67-hp engine of 900cc—the biggest boxer unit so far. Much more significant, however, was the influence of designer Hans Muth, who developed a smooth, integrated look, with an elegant dual seat and a sleek handlebar fairing giving a sporty cockpit feel. It was all impeccably turned out in shimmering metallic paint finishes and, at a stroke, BMW's "old man's bike" image vanished and a new, highly desirable identity was born.

BMW R100 RS and RT, 1976

With the 1976 R100 RS and its RT companion (left) came a new era in motorcycling—that of the luxury tourer, capable of sustained high-speed cruising all day with little fatigue for the rider and none for the engine. Intelligently cashing in on the success of the R90S, BMW developed an attractive full fairing in the wind tunnel to shield the rider and pillion passenger from the elements. The 70-hp engine, improved gear change, and rear disc brake helped the bike's agility on twisty sections, but the RS and more upright pannier-laden RT's real claim to fame was their perfect balance between the roles of long-distance tourer and sports bike.

BMW R45 and R65, 1978

After the success of the R90S and R100 RS had demonstrated a renewed demand for quality motorcycles, BMW planners felt confident to take on the successful Japanese middleweights with a new, lighter BMW model. With a smaller-capacity flat twin in a more compact frame, the R45 was aimed at insurance-conscious European markets, while the R65 (above) took over from the larger R60/5. Weighing little more than 200 kg, the bikes were compact, agile, attractive, and, of course, beautifully made. However, the price/performance balance was not competitive with Asian brands, and the neat BMWs struggled in the market. A late highlight was the 1982 R65LS, complete with wedge-shaped handlebar fairing that evoked the R90S.

BMW R80 G/S, 1980

As the 1970s drew to a close, BMW and its counterparts such as Triumph were facing a losing battle against the glossy, super-speedy Japanese machines. BMW knew it had to act and developed the four-cylinder K series, inspired by car technology, with the aim of phasing out the boxer engine. Nevertheless, the flat twin was given one last roll of the dice in an entirely new type of bike—a big-engined road bike that used experience from the Paris–Dakar overland rallies to tackle off-road trails too. With its 800cc engine, innovative single-sided swingarm, and adventurous image, the R80 G/S (below) not only saved the boxer engine and sired a massively successful line of new BMWs but inspired the entirely new adventure sport segment too.

43

5

War
and Postwar

For BMW—as for the rest of German industry—the war years were a period of astonishing pressure on all fronts. In the advancement of the war effort, engineering development, manufacturing technology, employment, and production planning were all pushed way beyond their familiar peacetime limits. In 1945, when it was all over, there was an appalling price to pay in terms of obliterated assets and, for many, a certain disgrace from being labeled an arms manufacturer and an employer of forced labor.

Yet in many ways BMW had it worse than its industry rivals. The Nazi war machine had forced it to completely realign its business, something that in prewar years had brought it spectacular success in terms of technical and industrial performance. However, it was precisely because of this success that the company had been all but destroyed—firstly by Allied bombing; secondly by the restrictions imposed by the American occupying forces; and, finally, by the way Germany had been parceled up into different zones by the victorious Allied powers.

Thus, by the summer of 1945, BMW found itself with all three sides of its business out of action. The Milbertshofen factory in Munich lay in ruins, and its machinery was scheduled for complete confiscation. So, too, was the Allach aero engine plant in the north of the city. There was a total ban on making anything remotely connected with aircraft, and motorcycle

∧ BMW's Milbertshofen plant was virtually destroyed by wartime Allied bombing, and occupying forces removed all machine tools, materials, and stores.

‹ Back to work: motorcycle manufacture restarted in 1948 with the single-cylinder R24.

45

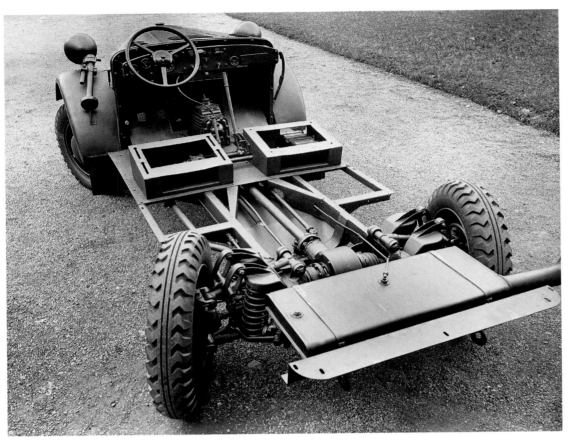

The 325 KFz3 built for the military was not a BMW design but featured an advanced chassis with drive and steering at all four wheels.

building was forbidden too—even in the unlikely event that materials could be found. Worse still, the entire car side of BMW's operations was inaccessible at Eisenach, now in the Soviet sector behind what would become the Iron Curtain; Eisenach housed the design offices, the engineering drawings, and the majority of the automotive expertise.

It was a humbling comedown after the intensity of the previous decade, which had seen BMW rise to become one of the world's leading aero engine makers as well as reach high points on two, three, and four wheels. It was these cutting-edge technologies that the Allied victors were so keen to seize when they marched into the BMW plant in 1945.

Dominant though the aero engine activities were, BMW was nevertheless an important wartime producer of terrestrial machinery too. Along with manufacturers Stoewer and Hanomag, BMW built the 325 KFz3—in effect, the German Jeep. The 325 was not designed by BMW but carried the company's six-cylinder engine (from the 326) and

had a sophisticated all-wheel-drive chassis, incorporating independent suspension and steering for all four wheels.

The BMW badge never appeared on the KFz3, but the roundel was worn with some pride by the company's military motorcycles, especially the R75 off-road sidecar combination (see page 39). This was the first attempt to build a sidecar outfit as an integrated whole, rather than simply attaching a chair to a solo motorcycle, and, most remarkable of all, the power was transmitted to the sidecar wheel, too. The combination's performance in extreme off-road conditions found favor with the Wehrmacht, and some sixteen thousand were delivered before air-raid bomb damage halted production.

Yet the remorseless pressure of aero engine building, imposed directly from Berlin, took its toll on BMW managers and the firm's determination to control its own affairs. Limited numbers of cars had been built until mid-1941, but by the following year the SS was pushing Franz Josef Popp to turn virtually the whole organization over to aero engine

production. This Popp refused to do, and he was promptly dismissed—with understandable fears for his safety. From then on, with Fritz Hille (who had joined with the Bramo takeover) in nominal charge, the SS ruled BMW much more directly, and the use of forced labor began to increase.

What could so easily have been the final call for the BMW concern came in April 1945 when the US Army marched into the Munich area and ordered the surrender or destruction of all BMW's machine tools, its stocks of components, and its design drawings. It was only thanks to good contacts between senior BMW managers and the US forces that dismemberment was avoided and permission received to return to a much more modest form of production: domestic pots and pans, bicycles, and even agricultural machinery.

This restarted the factory routine and, vitally, encouraged former employees to return. Before long, though in some secrecy, designs were taking shape for a 125cc two-stroke flat twin motorcycle—the largest that was permitted under the initial

∧ Overzealous marketing had led to the single-cylinder R24 being announced too early: the show example was a clever mockup.

BMW co-founder Franz Josef Popp was a key figure in the company until 1942, when he was dismissed by the SS for refusing to turn the whole operation over to aero-engine production. ∨

47

postwar restrictions—and BMW dealers across the country were being combed for spare parts for 250cc machines. The idea was to measure up every part to produce new working drawings for a fresh single-cylinder model ready for the anticipated raising of the capacity limit to 250cc, and by mid-1947 the design was all set. Much to the horror of the production engineers, who had no machine tools and knew full well that in the chaotic postwar economy materials were hard to find except by black-market barter, the R24 was unveiled at the Geneva Motor Show the following spring with a promise from the bullish export manager, Fritz Trösch, of deliveries before Christmas. Only a select few BMW officials knew that the elegant-looking prototype, with its shaft drive, telescopic forks, and smart unit-construction engine and gearbox, was in effect a dummy: It had no internal components, and some of the external engine parts had been expertly fashioned out of wood.

Miraculously, the promise of Christmas deliveries was kept, but only because of the major upheaval of the company going into receivership and having new management imposed by Deutsche Bank. The new president, Dr. Hans Karl von Mangoldt-Reiboldt, was adept at obtaining supplies and equipment and, aided by the improved security provided by June's currency reform and the new stabilized D-Mark, production was able to restart.

Motorcycles Lead the Way

The stylish R24 and its successor, the R25, were soon successful, and by 1950 almost thirty thousand had been built. Permission had now been granted for the manufacture of larger machines, and work on the R51 boxer twin was well advanced. Yet the profits from motorcycle sales were still too modest to fund a return to the car business, and, to make matters worse, there was still the bigger issue of BMW's Eisenach car plant, inaccessible in the Soviet zone. BMW Munich's anger was understandable: while it was itching to get back into car production, Eisenach was continuing to build and sell prewar BMW models—and many of these were finding buyers in what would soon be called "the West," right under the main company's nose.

The chosen solution to the problem was for BMW Munich to officially dissolve the Eisenach branch, making Munich the sole owner of the BMW trademark. Eisenach responded by relabeling its models EMW, but the dispute proved short-lived as the eastern plant later moved to produce new models under the Wartburg nameplate.

There was tension within BMW, too, over what sort of car should be produced once permission had been given. Attempts at shortcuts back into the car business, such as a tie-up with Ford or Simca or a merger with Auto Union, had come to nothing. And while engineering staff had proposed the small and elegant 331 two-seater coupe powered by the twin-cylinder air-cooled motorcycle engine, they were overruled by sales director Hans Grewenig and his plan for a return to the higher profit margins of the prestige sector. In this, Grewenig, who

▲ **The attractive 331 compact car was developed but never built—managers felt big cars would provide more profit.**

The 501, soon nicknamed the "Baroque Angel," was the first postwar car from BMW. The voluptuous Schimanowski-styled sedan had a new and very stiff chassis, double-wishbone torsion-bar independent front suspension, rack-and-pinion steering, and a novel central positioning of the gearbox. The 2-liter, six-cylinder engine was essentially a prewar design and gave just 65 hp, which struggled to overcome the model's 1,340kg mass.

1951: 501 prototype first shown in public.
1952: Production begins, with bodies built by Baur until BMW's in-house equipment is ready. High price levels hinder sales.
1954: V-8-powered 502 first announced. Few exterior differences, but advanced 100-hp all-alloy V-8 (the first in the world) gives much-improved performance. Lower-priced 501A and 501B double their sales.
1955: V-8 engine is enlarged to 3.2 liters and 120 hp.
1957: Uprated 3.2 Super gives 140 hp for a 177km/h top speed.
1959: Power steering and servo disc brakes added—the first on a German car.
1961: Range relabeled 2600 and 3200, with the 160-hp 3200S officially Germany's fastest sedan at 190km/h.
1963: Production ends, with 8,900 six-cylinder and 13,000 V-8 examples built.

had a banking background, echoed the upward move made by Popp in the 1930s.

The large and flamboyant 501 that emerged from this process in 1951 used the familiar six-cylinder engine from the prewar models but had completely new suspension, steering, and brakes, all attached to a massive box-section chassis. Though well received, the 501 was very expensive, heavy, and underpowered. The engineers were already aware of this, however, and had been working on a 2.6-liter V-8, a project that Fritz Fiedler was able to complete when he returned to the company from Bristol in 1952. The advanced all-alloy unit, giving 100 hp, was announced at the 1954 Geneva Motor Show; the 502 was visually identical to the 501 save for minor details. At the same time the 501 was reduced in price, and the result was a useful doubling of sales for 1954.

That same year saw the debut of a Mercedes-Benz model that would inspire BMW to produce one of the most celebrated designs in its entire history. Unveiled at the New York motor show, the 300SL sports car was a modernist sensation—it was directly derived from the firm's racing cars and its prodigious performance was matched by an equally high price. The 300SL idea had been encouraged by Max Hoffman, a charismatic Austrian émigré, who had been very successful importing European cars into the United States. With a perceptive eye for lucrative market segments, Hoffman

49

BMW 507, 1955–1959

The 507 two-seater roadster has become one of BMW's most famous models thanks to its gorgeous Goertz-inspired body, advanced V-8 engine, and impeccable hand-built construction. It has come to represent all the sporting and aesthetic values that BMW stands for, and it has taken on huge significance despite the tiny numbers sold during its short and deeply loss-making few years in production. Built on a new, shorter chassis with wishbone front suspension and improved location for the rear axle, the 507 enjoyed 150 hp from its high-compression V-8 with polished ports and a high-lift camshaft. In contrast to the 502 sedan and 503 coupe, the 507 had a direct floor shift. However, hampered by high prices, it never sold in the hoped-for numbers.

1955: Debut at IAA show in Frankfurt.
1956: First deliveries begin.
1959: Disc-brake version shown; production halted in December with just 252 examples made.

▲ **The success of the sensational Mercedes 300SL prompted BMW to build the 507.**

suggested to BMW managers that they, too, should consider a model targeted at the smart and fashionable of the US East Coast and California; however, he counseled that the car should be positioned midway between the impossibly expensive Mercedes and the cheap and cheerful British Triumphs and MGs that most American sports enthusiasts were buying.

Thus the remarkable BMW 507 was born. It was a happy amalgam of the 502's V-8, by now uprated to 3.2 liters and 150 hp; a stiff, shortened chassis; and a two-seater roadster body of stunning proportions. The now-classic style had been penned in record time by Count Albrecht Goertz, another émigré German in Hoffman's circle of contacts, after in-house designs had been rejected by Hoffman—though it is not clear whether these were for the 507 or the longer and less sporty 503 four-seater developed in parallel. Goertz had worked with the notable Raymond Loewy on Studebakers and would later be hired as a consultant for Nissan on its Z-series sports coupes. Legend has it that the board hurriedly approved the project while company chairman Kurt Donath, who opposed the 507 because he felt BMW's priorities should be elsewhere, was in hospital.

The celebrated 1955 507 roadster; its designer, Albrecht Goertz; and (top) the 1999 Z8 it inspired.

The four-seater 503 coupe and cabriolet were less glamorous than the 507 but sold more steadily. ❯

Donath did have a good point, however: BMW was losing money fast and needed quick-selling smaller cars rather than indulgent image-building models for the wealthy. In time his view was vindicated—in the decades since its 1956 debut, the 507 may have become one of the most cherished icons of world automotive design, but in the harsh reality of the 1950s, it was a commercial disaster that helped bring BMW to the brink of bankruptcy.

Hoffman had forecast that he could sell some five thousand units a year in the United States at a price of $5,000 each. But such was the cost of production in Germany that the launch price swelled to more than $9,000—more than the 300SL's—and later rose well into five figures. Demand proved to be low, and even at that high price the company lost money on each example built. To the probable relief of BMW's accountants, but not sports enthusiasts, the company terminated production in March 1959 after just 252 examples had rolled off the Munich lines. The less racy and less familiar 503, which paradoxically sold almost twice as many, was withdrawn at the same time. Even though the 507 failed on a commercial level, it did succeed—albeit briefly—in its main aim of putting BMW back on the sports-car map and signaling the talent of the company's engineers.

The 507 proved to be an extraordinary car on many levels, not least of which was an enduring influence on the

∧ BMW's range in the mid-1950s offered both extravagant luxury and back-to-basics bubble-car motoring, but nothing for ordinary family motorists. In the company museum a 502 cabriolet sits alongside a 502 chassis and an Isetta bubble car.

world of automotive design out of all proportion to the tiny numbers actually sold. It is now seen as a pivotal design in the BMW story and has been referenced with some regularity in later sports models such as the Z3, the Z07 concept, and the Z8. Its timeless proportions make it a habitual favorite of those who compile top-ten automotive icons lists, and it regularly strays into six-figure territory on the rare occasions one of the survivors comes up for auction.

If perseverance with the 503, the 507, and ever more powerful versions of the V-8 502 might have appeared like delusions of grandeur on the part of the BMW board, then the abortive 505 project provided the confirmation. Appearing at the same 1955 Frankfurt motor show as the 507, the 505 was an ambitious attempt to challenge Mercedes on its private turf—the grand limousine for heads of state, diplomats, and corporate magnates. With formal and upright styling by Giovanni Michelotti, the 505 got off to a poor start on BMW's Frankfurt show stand when, according to industry folklore, German chancellor Konrad Adenauer knocked his hat off when getting into the back seat. This public embarrassment effectively canned the 505, ensuring that the Mercedes 300 retained its honorific title of the Adenauer Mercedes.

Undaunted, the urge to take on Mercedes at its luxury game remained a strong driving force at BMW. Two decades later, with the arrival of the 7-Series, the challenge was for real, and by the 1980s BMW had drawn level and was poised to pull ahead.

Yet for every grandiose scheme put forward by BMW managers, there was a more grounded faction intent on finding more realistic solutions to the company's woes. As early as 1954, statisticians had begun to spot an easing in demand for motorcycles as Germany's *Wirtschaftswunder* began to gain pace and buyers raised their sights from two wheels

BMW 503, 1955–1960

Also styled by Goertz, the 503 was the often overlooked, less glamorous sister to the universally celebrated 507. Built on a longer chassis to allow four-seat accommodation, it shared much of the 507's technical specification but sold more consistently, despite an even higher price. Particularly prized was the cabriolet version, one of the first European soft tops with an electrically powered hood.

1955: Debut at IAA Show in Frankfurt.
1956: First customer deliveries.
1960: Production ceases.

❮ An attempt to challenge Mercedes, the 505 limousine
did not get past the prototype stage.

The egg-like Isetta sold well and helped stave off BMW's collapse.

Heinrich Richter-Brohm took charge in 1956 with the mission of developing a new midsize car, but the funding was not forthcoming until after 1960.

to four. Realizing that this was not a blip but a worrying trend, BMW approached Iso in Italy, whose tiny Isetta bubble car had become popular in its home market, with a view to taking out a licence to build the car in Germany and equip it with BMW's single-cylinder motorcycle engine. Iso owner Count Renzo Rivolta was happy to agree, and things proceeded so swiftly that the prototype appeared at the Geneva show within a matter of months.

Production was soon underway, and the Isetta proved an immediate success, selling almost thirteen thousand in 1955 and triple that amount in the two following years. By 1958, with sales pushing into six figures, it would become the biggest seller the company had ever made. Yet even that extra lifeline was not enough to compensate for the dwindling revenue from motorcycles, let alone the continuing losses from the luxury cars. BMW was an automaker with no products in the middle of the market, where all the growth and profit resided, and something had to give.

Deutsche Bank had been a loyal and increasingly big shareholder in the company, and in the 1930s it had helped provide the impetus to move into the car sector. By 1956 its representative on the BMW board, Heinrich Richter-Brohm, had seen enough worrying signs that he precipitated a management reshuffle, becoming managing director. On appointment he asked to see the company's future plan and was shocked when informed there was none. Richter-Brohm's response was to draw up a 133-page report analyzing the situation and suggesting suitable solutions. "This company must change from the ground up," he advised, prescribing a completely new midrange BMW car to be produced at twenty-four thousand units a year to generate a turnover of DM 300 million and a balance-sheet profit of DM 14 million. The vehicle was slated for launch in 1959, and by the company's annual meeting in late 1957, a super-confident Richter-Brohm claimed that the finance was in place "for the new production programme designed to fill the gap between the Isetta and the big car."

Few companies have ever had such a yawning gap to fill, and yet despite the desperate urgency of the situation, Richter-Brohm's finance failed to fully materialize and only an interim project—the BMW 600 "big-bubble" car—became reality. Though the 600 sold well in its first year, it quickly collapsed, and cash flow once again reduced to a trickle. A flicker of hope appeared at September 1959's Frankfurt show in the stylish shape of the Michelotti-penned 700 coupe, based on the platform of the 600—orders duly flooded in, but it would be too late to stave off the threat of insolvency.

By the December shareholders' meeting, the vultures, including Mercedes-Benz, were already circling, and the game seemed to be over. Indeed, it was—until an adjournment was called when the meeting became too riotous. The lawyers found a fatal flaw in the board's figures, which enabled a hitherto low-profile shareholder, Herbert Quandt, to step in

and snap up the unwanted stakes to swing the vote in favor of remaining independent.

Had it not been for Quandt's decisive action, BMW would have become a tiny token brand in the Mercedes empire or, worse, wound up completely. But the fresh start and the new vision supplied by Quandt were just what BMW needed, and they would provide the inspiration in the coming decades to propel the Munich company ahead of its archrival to become the world's leading producer of premium cars.

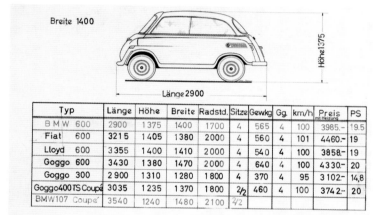

Typ	Länge	Höhe	Breite	Radstd.	Sitze	Gewkg	Gg.	km/h	Preis mt Heizung	PS
B M W 600	2900	1375	1400	1700	4	565	4	100	3985.–	19.5
Fiat 600	3215	1405	1380	2000	4	560	4	101	4460.–	19
Lloyd 600	3355	1400	1410	2000	4	540	4	100	3858.–	19
Goggo 600	3430	1380	1470	2000	4	640	4	100	4330.–	20
Goggo 300	2900	1310	1280	1800	4	370	4	95	3102.–	14,8
Goggo 400 TS Coupé	3035	1235	1370	1800	2/2	460	4	100	374 2.–	20
BMW 107 Coupé	3540	1240	1480	2100	4/2					

⌃ As a grown-up Isetta, the BMW 600 had a brief flurry of success but faded fast.

The smart 700 minicar was designed around the hardware of the 600. ⌄

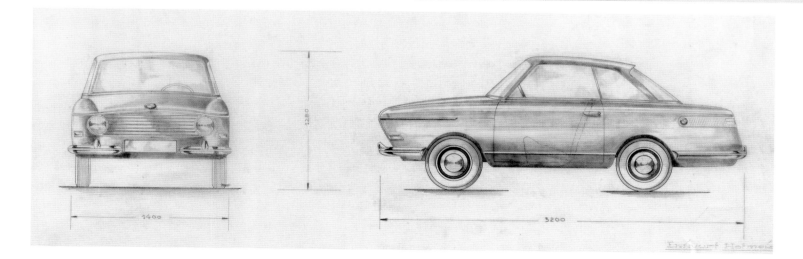

6

Thinking Small

BMW embarked on its carmaking career in 1928, when times were hard and when sales of large models had stagnated. The company's choice of an off-the-shelf small-car design proved a shrewd one, and BMW was able to build on the success of the Dixi to move to larger and more profitable models. Yet the streak of ingenuity required for successful small-car design has always been present within BMW and has returned to the fore whenever strategy demanded, whether it was the bubble cars for the 1950s fuel shortages, the classy miniature coupes in the 1960s, or the first premium small car in the shape of the 2001 Mini Cooper.

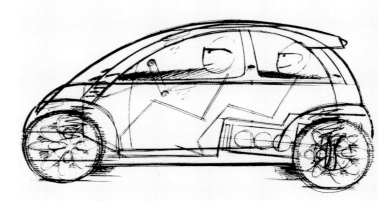

Dixi DA1, 1928

From its very first days as an aero engine producer and motorcycle maker, BMW had harbored the long-term ambition to move into the car business. After a series of internal and technically advanced prototypes had failed to pass muster and BMW management was locked in indecision, the company jumped at the opportunity to build the British Austin 7 under license, taking over the Dixi works and its manufacturing contract for the tiny four-seater. BMW systematically refined the lightweight design, improving braking, steering, suspension, and quality until it became a desirable small car. This experience paved the way for later iterations that owed less to Austin's cost cutting and more to BMW's engineering skills, with the 3/15 derivative gaining the BMW roundel badge and the precursor to the double-kidney grille.

BMW 331, 1950

Many regret that the shapely 331 (above) never passed the prototype stage. Conceived by chief engineer Alfred Böning in the late 1940s amid cash and materials shortages and an uncertain future, the 331, also known as the 531, made sound commercial sense—it used the existing 750cc flat twin motorcycle engine, front mounted and driving the rear wheels through a four-speed gearbox. While its two-seater coupe format would have limited its appeal, its elegant Topolino-like lines might have made it a minor classic. The 331 could have been BMW's first postwar production car, but the project was scrapped in favor of the greater potential profits at the opposite end of the market, with the extravagant luxury-segment 501 Baroque Angel series.

BMW Isetta, 1955

Conscious that its mainstay motorcycle sales were slowing as rising German prosperity in the 1950s encouraged buyers to aspire to the greater comfort and status of four wheels, BMW began combing Europe for solutions. Count Rivolta in Italy was happy to provide the answer in the shape of his egglike Isetta bubble car (right), which BMW re-engined with its single-cylinder 250cc four-stroke from the R25 bike. With

Italian Isetta sales being hurt by the new Fiat 500, Rivolta was keen to sell BMW the tooling too. It proved to be money well spent, for the Isetta was an instant hit, selling strongly through to 1958 and providing BMW with its best showroom performer to date. Intelligent updates, including a bigger engine and sliding side windows, kept the Isetta fresh, but by the turn of the decade, the bubble car's days were over and it bowed out gracefully in the knowledge it had helped keep BMW afloat through tough times.

BMW 600, 1957

BMW's financial woes in the mid-1950s were in no small part attributable to the yawning chasm between the tiny Isetta bubble car and the slow-selling, heavily loss-making luxury models. Lack of cash ruled out the fresh middle-class car for newly mobile families that the company really needed, so managers decided as a stopgap solution to extend the bubble-car concept with the much larger 600 model (top). Seating four and featuring a side door in addition to the opening front door—with steering wheel still attached—the 600 used the firm's twin-cylinder fan-cooled motorcycle engine, tucked away in the tail. After a brief flutter of success in 1958, the 600 disappeared almost without trace, the victim of its ungainly microbus looks. Today it is remembered mainly as a marketing curiosity, but also as the first car to feature semi-trailing-arm rear suspension—the chassis feature that would later give BMWs the handling edge over their Mercedes rivals.

BMW 700, 1959

The story of the BMW 700 (above right) is, uncharacteristically for BMW, one of corporate dissent, division, and intrigue. Frustrated by the big-bubble 600's lukewarm reception and by the refusal of its designer, Willy Black, to change tack,

managing director Richter-Brohm responded to an offer by his Austrian agent Wolfgang Denzel to produce a more attractive design using the same mechanical elements. Working in Vienna with Italian stylist Michelotti in conditions akin to those of a secret skunk works, Denzel mapped out the design of what would become the 700 and gained Richter-Brohm's approval. Midway through 1958 he unveiled the prototype at BMW's Lake Starnberg hideout, and the response to the freshly styled, clean-lined design was instantly ecstatic. The following year

the 700 appeared in sedan and coupe form at the Frankfurt show, triggering a rush of advanced orders. The design, which incidentally also featured BMW's first monocoque construction, survived the firm's near-bankruptcy in 1959 and went on to become a big seller in the marketplace as well as a strong performer on the motorsport scene.

BMW E1 (Z11), 1991

Once BMW's sales and its performance image had begun to take off in the 1960s, there was little incentive to return to small cars in order to tempt additional customers. But with growing environmental awareness and the prospect of a mandatory quota of zero-emission cars in California in the 1990s, BMW began working on concepts for battery-powered city cars. Much earlier, the company had developed an electric version of the 1602 for the 1972 Summer Olympics in Munich. The tiny E1 city car (above left), presented in 1991, used an aluminum-and-composite structure, seated four—plus luggage—and claimed a perhaps optimistic 120km/h top speed and 200km

range from its sodium-sulphur battery mounted under the rear seat. The E1, so called to stress its pure electric driveline, was fully roadworthy and is cited by BMW as an important steppingstone in the engineering progression that led to the i3 megacity car in 2013.

BMW Z13, 1993

With this low-slung city commuter (opposite, bottom), the first signs of BMW's vision of future megacity vehicles become evident—a line of thinking that twenty years later would result in the revolutionary all-electric i3. Though the Z13's ultra-low front and goldfish-bowl windscreen now look dated, in its layout it anticipated many modern developments. The four-cylinder motorcycle engine at the rear echoes the arrangement of today's Smart and Renault Twingo, and the three-seater layout, with a central driving position, is precisely that specified by Gordon Murray—both for his remarkable BMW-powered McLaren F1 supercar, first shown the year before, and in his recent series of city-car concepts employing his radical iStream construction. To suit its premium aspirations, the Z13 was packed with luxury equipment, even including an in-car

fax machine. With its lightweight-aluminum structure, the Z13 promised excellent economy for its time. However, that time never came, as BMW's takeover of small-car specialist Rover the following year signaled an end to any production plans.

Mini Cooper, 2001

The Mini brand was one of the most prized assets that came with BMW's 1994 purchase of the Rover group, and the two companies quickly began debating what the modern replacement for the classic 1959 Mini should be. Some factions backed a radical rear or underfloor-engined approach as the most true to the Mini spirit of innovation; others argued that the Mini character lay in the look, feel, and fun handling of the original. The Mini prototype that emerged from this prolonged process in 1999 reflected the latter view, and though it did indeed convey the classic model's visual cues with great authenticity and affection, it had grown substantially in bulk and was no longer so space efficient. But as the first premium small car, the BMW-inspired Mini was a massive hit from the moment it was launched in 2001 and has gone on to change the way the whole world regards high-class small cars.

7

The Quandt Years

T he dramatic events of the shareholders' meeting on December 9, 1959, have been frequently and fully documented in the many accounts of that pivotal day: the ultimatum from the pro-Mercedes block; the discovery of a possibly deliberate flaw in the corporate bookkeeping, allowing an adjournment; and the emergence of a blocking minority with a well-funded backer—the Quandt organization—as well as the loyal, small shareholders and the dealer owners.

Yet behind the euphoria over the company being saved from the clutches of Mercedes-Benz lay the harsh reality that this was still a business in deep trouble. Its overheads were crushing, its model policy—split between bubble cars and outdated luxury cars, both of them declining segments—was without any logic, and motorcycle sales were slowing too. The only bright spot was the tiny 700-series coupe that had just secured an encouraging advance order bank.

Herbert Quandt and his half brother, Harald, were contrasting personalities. Contemporaries painted Herbert as secretive but willing to be adventurous in business, while Harald—whose stepfather was Nazi propaganda chief Joseph Goebbels—was gregarious but more careful with his money. Yet neither took charge at BMW, nor ever accepted a seat on

^ **The brand-new BMW 1500 pulls in the crowds on its 1961 show debut.**

63

the board. Instead, they set about drawing in the necessary management talent from outside. Herbert was especially adept at gaining the confidence of banks and attracting the right people; before long he had hired chief engineer Wilhelm Heinrich Gieschen from the struggling manufacturer Borgward and poached sales manager Paul Hahnemann, who had been just about to leave Auto Union.

Also on board was Gerhard Wilcke, who had been Quandt's legal adviser for several years. He was particularly skillful in handling the relationship with Daimler-Benz, where Herbert still had a stake and where many believed that a hostile takeover was still a threat. Wilcke was also able to deal with approaches for BMW from companies such as American Motors (which had hoped to build its Rambler in Munich), Chrysler, Simca, Britain's Rootes Group, Fiat, and Ford.

What Quandt provided was more than just the confidence of the banks. He quickly formed a clear vision of BMW's ambitions for the future, where the company aimed to

> ❯ **Herbert Quandt, right, was a hands-off majority shareholder in BMW but was very astute at hiring the right people for key roles.**

be in terms of image and products in five or ten years' time. His stewardship provided the stability that allowed BMW's pent-up engineering talent to blossom once more, as Alex von Falkenhausen, racing driver turned engine-design genius, testified in Eric Dymock's *BMW: A Celebration*:

> *"I think the main problem was that every year we seemed to have a new board of directors," explained von Falkenhausen. "They changed very quickly. It was difficult to come into contact with a new director—they were all in despair. But the people under the directors wanted to do something, so they tried very hard."*

Herbert Quandt had known all along that the most important priority for BMW was to develop and launch the much talked about midsize car. Every effort was channeled into the so-called "Neue Klasse," or new class. Quandt was able to muster the necessary resources in short order, aided by the timely collapse of Borgward, a producer of quality midrange cars in a similar segment to BMW's target market. This not only removed a potential competitor from the marketplace but also released a coterie of engineers experienced in just the areas BMW needed—production engineering, testing, and quality control.

Quandt had set an almost impossibly tight development schedule, yet, astonishingly, the combined talents of BMW and ex-Borgward engineers succeeded in completing a prototype in time for the Frankfurt show in September 1961—the car was unveiled as the BMW 1500. The newcomer, a four-door sedan measuring 4.5m in length, was fresh and attractive in its looks, penned by Wilhelm Hofmeister with Italian design studio Bertone reputedly in the background as consultants. The design immediately won praise for its style and modern image; the engine, meanwhile, impressed technical commentators with its overhead camshaft and claimed output of 75 hp, at that time a very good figure for a 1.5-liter.

BMW, too, was conscious of the significance of this comeback model. The company's press material for the 1500's debut described the event in grandiose terms: "In the history

of automobile construction there are few models which have been the subject of so much public interest and discussion as the new medium-sized BMW." Warming to its theme, the press release went on to laud the 1500's "sportiness, refinement, genuine comfort and timeless elegance," noting that the mission of the 1500 was to bring the qualities of BMW's big V-8s into a smaller-capacity car, and that BMWs, in general, had a tradition of being ahead of their time.

The 1500 weighed in at just 900kg, claimed BMW, which allowed the company to get away with no servo assistance for the disc front and drum rear brakes; likewise, power assistance for steering was unknown except on top luxury cars. The interior, though stark by today's standards, was seen as stylish

and attractive, and safety features included mounting points for seat belts for all four occupants, a padded steering wheel boss, and recessed instruments—just speedometer, clock, temperature, and fuel level.

BMW's claim in press materials that the M10 engine "would be up to the minute for the next ten years" was not just an idle boast. Von Falkenhausen's design, which married an iron block with the aluminum head containing the chain-driven overhead camshaft, was deliberately conceived to be capable of expansion up to 2 liters in capacity. This did, of course, soon happen, and the M10 remained in production until 1988, with more than 3.2 million made. It powered not just the Neue Klasse cars but also the '02 series, including the infamous Turbo; two generations of the 3- and 5-Series; and, most remarkably of all, BMW's first Formula One effort, where it yielded up to 1,300 hp in highly turbocharged qualifying trim.

The company had high expectations of the 1500, but even the most optimistic of its managers were taken aback by the enthusiasm with which the public took to the new car. Even the high price of DM 9,485 appeared acceptable to the customers who eagerly signed up for the waiting list. Paul Hahnemann declared at the Frankfurt show launch that that price included all necessary equipment, such as disc brakes, heater, two-speed wipers, and windscreen washer, and that it left "little room for profit." Only higher volumes could boost profits, he predicted.

Hahnemann, in concert with a marketing psychologist, had been one of the first auto industry figures to come up with the idea of niches. BMW, he said, needed to aim for the same combination of compactness, sportiness, and quality that had served it so well in the 1930s. Priced at just under DM 9,500, the 1500 was, he said, specially targeted at the rising market of middle-class buyers. He went on to list its competitors as the

The 1500 established a BMW style that would last more than twenty years, its eager engine even longer. The dashboard was seen as a model of clarity and tasteful design.

Opel Rekord, Peugeot 404, Taunus 17M TS, and Volvo Amazon at the lower end and the Mercedes 180 and 190, Citroën ID19, and Opel Kapitän in the more expensive class.

Eberhard von Kuenheim, yet to begin his astonishing twenty-four-year tenure at the helm of BMW, recounted in a 2003 interview with the author how Herbert Quandt deliberately encouraged the premium price policy at that time:

Yes, the price was high but it was accepted by people who said, "I can afford that. I won't drive a Beetle anymore—everybody has a Beetle." The effect of this was that BMWs became the cars for "Aufsteiger"—the social climbers of the era, rising up the ladder of social status. The 1500 series was doing well, and there was also the small two-door car, the 1600, 1800, and then the 2002.

Perhaps predictably, those who had signed up for the born-again BMW had to wait rather longer for their cars than they had expected. It would be October of the following year before the first customer cars were delivered, yet everyone agreed that the 1500 was worth the wait—and the occasional gremlin. The engine, now boosted to 80 hp in production trim, gave lively performance on the road, and the independent semi-trailing-arm rear suspension, developed from that of the 700 coupe, provided responsive handling and a good ride. Quickly, the 1500 gained a reputation as *the* car for the sporting driver to have.

The unexpected scale of the 1500's success did, however, present BMW managers with some awkward problems. The most acute of these was the inability of the company's new production facilities to keep up with demand for the new car—this was new territory, as BMW had not

BMW 1500–2000, 1961–1972

The Neue Klasse, or new class, was the much-needed midsize car that had been in BMW's product plans since the mid-1950s but which was only developed after the rescue of the company in 1960 brought the necessary funding. Fresh in its style and engineering, with a potent overhead cam engine and independent rear suspension, the four-door 1500 represented a return to BMW's heartland of compact, sporty cars and was the springboard for the later 3-, 5-, and 7-Series sedans and BMW's rise to market leadership.

1961: Prototype 1500 unveiled at Frankfurt show.
1962: First production models delivered, with engine power raised to 80 hp.
1963: 1800 model added, with 90 hp.
1964: 1500 becomes 1600, and 1800 TI launched with 110 hp or as high-performance 1800 TI/SA, a short run of two hundred cars sold only to drivers with a racing licence. The engine gave 130 hp with full inlet and exhaust silencing in place.
1966: New 2000 featuring 2-liter engine from 2000 coupe offered with 100 and 120 hp, giving top speeds of 168 and 180km/h.
1968: Revised 1800 gets a new engine with 2-liter block and pistons but 1600's short-stroke crankshaft.
1969: Launch of 2000 tii with Kugelfischer mechanical fuel-injection system.
1972: Production ceases in preparation for 5-Series, with a total of 350,000 made.

BMW 1500 — Die neue Klasse

had such a popular model for a generation or more. It was thus that in order to free up more production capacity for the Neue Klasse, the tough decision was taken to axe two models of great sentimental importance but, at best, marginal profitability: the Isetta and the long-serving Baroque Angel limousine. The small but smart 700 sedan and coupe, by contrast, were retained as they were selling well, and demand had strengthened with the launch of the lengthened LS versions with their much-improved interior space.

Confident that the 1500 had hit the sweet spot of the market, Hahnemann began applying another successful policy from the 1930s—developing a wider range using variations on the same set of components. Now, clearly, was the ideal opportunity to further exploit the market it had just created, and derivatives soon followed.

The 1800 offered a substantial power hike to 90 hp, but more importantly it also provided the basis for the 110-hp 1800 TI, first in a long line of derivatives designed to appeal to enthusiastic and sporty drivers. Listed as extras for the TI were competition equipment including stiffer springs, a five-speed gearbox, a selection of axle ratios, competition brake linings, a 105-liter fuel tank, and various engine components for still more power.

"The 1800 TI is a car for exacting people who drive it for the sake of driving," noted BMW. But soon, in its turn, the already special TI prompted an even more special edition still. The 130-hp 1800 TI/SA, whose name stood for "SportAusführung," or sports version, was a short run of two hundred cars sold only to drivers in possession of a racing licence. Despite its intimidating price of DM 13,500, the 180km/h-plus TI/SA quickly became must-have equipment for touring car racers in Germany and across Europe and was the first of many high-performance BMW sedans to set both race and road driving alight. Arguably, the TI/SA and the later tii paved the way for the M versions of subsequent model generations.

A standard 2-liter version of the M10 engine already had appeared in the 2000 CS coupe, an elegant 2+2 on the Neue Klasse platform styled by Manfred Rennen under Hofmeister but influenced by Bertone's earlier 3200 CS, the final fling of the V-8 and the old 502 chassis. It was not long before BMW applied the bigger four-cylinder to the four-door sedan to move the range still further up the price and performance scale.

^ The shorter, two-door 1600/2 was launched on BMW's fiftieth birthday outside Munich's opera house. Its runaway success as the '02 series astonished even BMW.

BMW claimed the 120-hp 2000 TI was the "absolute fastest production car at the Nürburgring," and this model quickly became the fast sports sedan to be seen in. Yet, towards the end of 1969, BMW went one better and fitted fuel injection—the first time the firm had employed it on a road car—to the 2000, creating the tii model, a designation that was to become even more famous on the compact 2002, launched eighteen months later.

The fuel injection—at that time an exotic novelty fitted only to a few top sports and luxury cars—gave the 2-liter an extra 10 horsepower over the twin-carburetor TI, boosting top speed to 185km/h and making for something with no real competition in the car market. Buyers had to pay quite a significant premium for the tii version, ensuring that sales only ran at a small fraction of those of the standard 2000; the biggest seller of all remained the 1800.

A Long-Term Plan Emerges

On the industrial front it had been a tale of gathering momentum. Between 1960 and 1969 production and employment had tripled, while turnover had increased sixfold. Genuine long-term planning had begun in 1965 when BMW bought the Glas company in Dingolfing, as much for its production capacity and spare land as for its pretty coupes with their innovative engines. In motorsports the 1800 TI won the Spa 24 Hours in 1964 and again the following year, and establishment of a proper international dealer network began.

Sensing the good times were here to stay, Hahnemann wanted more. The grand plan was to expand both upward and downward, the former by building a six-cylinder engine and putting it into a larger car to fire a warning shot at Mercedes, and the latter—seen as the more cautious of the two moves—by producing a simplified, cut-down, two-door version of the core car to sell at a lower cost.

That car was, of course, the now-legendary '02 series, which began as the 1600-2 (indicating two doors) in 1966 and which succeeded even beyond BMW's wildest expectations. The shortened body style, created by George Bertram, initially appeared somewhat unbalanced, but with 85 hp and a weight of under 940kg the 1600 was a real flier.

Right from the start buyers couldn't get enough of these "beginner" BMWs. In its first full year on sale, the two-door straight away matched its bigger brother in the showroom; year two saw it sell twice as many, while after three years, at almost eighty-five thousand units, it more than quadrupled the bigger car's sales performance. Americans took to the baby BMW with particular relish. In an era when Detroit's products were still overweight, oversize, and dull to drive, the quick and accurate responses and the fresh, energetic powertrain of this modest machine came across as an absolute revelation. Suddenly, Americans discovered that driving could be fun and not cost the earth, and that German products with quality and integrity were more affordable than they had imagined. *Car and Driver* splashed the 1600-2 on its front cover, loudly proclaiming it "the world's best $2,500 automobile."

More powerful versions soon followed, and with the 2002 tii BMW straight away hit the big time. This model was

CHAPTER SEVEN

24ᵘ van Francorchamps sensationele overwinning van de BMW 1800 TI

1ᵈᵗᵉ in zijn categorie tegen het geweldige uurgemiddelde van 164 km

↑ **Race winner: the 1800 TI was a regular victor in endurance racing.**

BMW 1602–2002, 1966–1977

Originally intended as a lower-cost, cut-down alternative to the Neue Klasse sedan, the '02 series was lighter and just as powerful and became a dramatic success. The most sporting 2002 tii helped establish the BMW name among keen drivers, while the Turbo's launch was poorly timed and the unexplained failure of the Touring hatchback stands as one of BMW's few marketing misjudgements.

1966: 1600-2 presented on BMW's fiftieth birthday; 85 hp and a curb weight of 940kg give it strong sporting appeal.
1967: 1600 ti launched, with 105 hp.
1968: 2002 launched, with 100 hp; 2002 ti follows, with twin carburetors and 120 hp. Cabrio, built by Baur, launched. Replaced by in-house cabrio with rollover hoop in 1971.
1971: 1602 Touring with hatchback tailgate and folding rear seats, later available with larger engine options; dropped in 1974 because of slow sales.
1971: 2002 tii with fuel injection and 130 hp becomes an instant hit; 1802 added, with 90 hp.
1973: Launch of 2002 Turbo, with mechanical fuel injection, turbocharger, and 170 hp, giving 211km/h top speed; dropped in 1975 amid fuel-crisis pressure.
1975: 1502 runout model with 75-hp economy engine is an unexpected success and continues until 1977, overlapping with its successor 3-Series; more than 830,000 '02 models would be built in all.

unquestionably the most thrilling small car on sale anywhere, and the world soon sat up and took notice. Among car enthusiasts the buzz was no longer Alfa or Porsche or Jaguar. The car for the quick driver to have—even the very rich quick driver who could afford to park a Porsche alongside the Mercedes in his garage—was a tii. As an evocation of BMW brand values, the tii was perfect, an emblem of everything that was fast and fun about driving.

The '02 went on to spawn three more derivatives, the first two a convertible and the Touring hatchback; the Touring was one of the rare instances where Hahnemann's niche theory failed to ignite the expected spark among buyers. The most extreme derivative, however, was the 2002 Turbo with an explosive 170-hp engine taken from the futuristic 1972 Turbo concept car. Though it generated great excitement among enthusiasts, it was tricky to drive and came just as the global fuel crisis struck. BMW, sensing the Turbo was inappropriate for the times, quietly withdrew it after just 1,672 had been built.

At the opposite end of the scale, BMW's six-cylinder limousines did not have the easy ride enjoyed by the smaller cars. The 2500 and 2800 were deliberately designed by

BMW 2500 and 2800, 1968–1977

BMW's return to the six-cylinder market in 1968 saw it square up to Mercedes-Benz with a full-sized luxury car reflecting the brand's sporty driving character. High perceived prices and firm suspension limited the models' appeal on international markets, but performance and specification were systematically improved over the design's ten years in production.

1968: Presentation of 2500 and 2800 sedans, with 150 and 170 hp. Disc brakes and independent suspension on all four wheels, optional automatic transmission, power steering, and limited slip differential; first BMW to offer trunk-lid mounted toolkit.

1971: Bavaria special for the United States, with larger engine and simpler specification, helps energize slow sales; 2.8-liter engine enlarged to 3 liters and 200 hp.

1973: Launch of 3.3-liter model with extended wheelbase and 190-hp carburetor engine.

1976: US-compliant 3.3-liter with fuel injection and 197 hp.

1977: All models cease production, with total standing at 222,000.

The elegant 2000 CS coupe was based on Neue Klasse mechanicals but grew a longer hood to accept six-cylinder engines, later spawning the legendary CSL.

Hahnemann to feel responsive and sporty in a class where soft suspension and soggy steering were the norm, and initially BMW officials delighted in press and customer reports of eager and super-smooth engines and a firm ride. These comments, reasoned Hahnemann, reinforced BMW's sporting image. However, the mix was even more poorly received in North America, the principal target market, and, compounded by adverse currency movements, the big sixes struggled in most of their territories and failed to hit sales targets, leading to stockpiles of unsold cars.

By this time BMW had other problems too. Hit by the rise of the Japanese competitors, motorcycle sales had dropped to under five thousand units in 1969, prompting some to suggest a complete withdrawal from the two-wheeler market. Instead, the decision was made to move the whole

▲ One of the first turbocharged cars on the market, the 2002 Turbo was fast but tricky; it was withdrawn in deference to the 1973 fuel crisis.

▼ The large 2500 and 2800 sedans, with their super-smooth six-cylinder engines, brought BMW back into the luxury segment.

▲ The 1968 2800 CS evolved into the potent 3.0 CSL, which became legendary in the world of motorsport.

bike operation to Spandau in Berlin, and the launch of the fresh /5 range helped revive interest.

More tragically, Harald Quandt had died in a light-aircraft accident in late 1967, and speculation once again arose that BMW might become a takeover target. Several potential buyers did indeed put forward offers, but Herbert Quandt swiftly batted them away. The real crisis, such as it was, surfaced in 1969 after Gerhard Wilcke resigned as president. Sales director Paul Hahnemann, who many took as the man in charge because of his big personality, expected he would be named as Wilcke's replacement. Instead, it was a much younger aide to Harald Quandt, a Prussian aristocrat by the name of Eberhard von Kuenheim, who was given the top job. His was a name that would figure centrally in BMW's history for more than a quarter of a century.

FLIGHTS OF FANCY

BMW's Concept Cars

As a company confident of its identity and its direction, BMW has rarely had to resort to concept cars to broadcast engineering achievements or to test public opinion. Yet highlights such as the 1972 Turbo and 2009 Vision EfficientDynamics have signaled the start of whole new ways of thinking for the automotive business, and the Hommage series of design studies pay exquisite tribute to the historic models that made BMW what it is today.

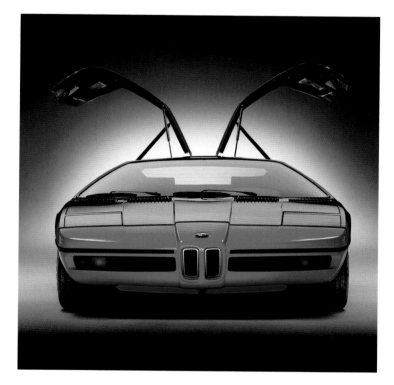

Turbo, 1972

With the announcement that the 1972 Summer Olympic Games would be staged in its home city of Munich, BMW knew it would have the attention of the whole world for a few precious weeks. A high-profile new vehicle was prescribed, and chief designer Paul Bracq duly delivered the goods with the stunning Turbo, a two-seater mini-supercar that established a design language for extreme sports cars that would prove highly influential for many years to come. It would also directly spawn Giorgio Giugiaro's equally stunning M1 in 1978.

The Turbo also revived the idea of gullwing doors, citing an English patent of 1938, and the feature would become a must-have for every supercar. Yet surprisingly, despite its breathtakingly modern, low, wide, and flat proportions and the then-impressive 280-hp potency of its 2-liter, eight-valve engine, the Turbo was billed by BMW at the time as a rolling research laboratory. Its proclaimed role was to test out active

75

and passive safety features that went way beyond the then-prevailing US norms—hydraulic bumpers protected the front and rear from damage; the seat belts were interlocked with the ignition to prevent unbelted travel; and a radar distance warning automatically backed off the throttle if the driver got too close to the traffic ahead. It took a further thirty years before radar cruise control became available on production cars, but Bracq's sensational style would make its impact much sooner and much more widely.

Z21, 1995

Something of a historical and engineering curiosity, the Z21 (below right) was a 1995 study by BMW Technik GmbH for a minimalist two-seater sports car with similar appeal to a Caterham or, more recently, KTM's X-Bow. As on those track-focused specials, the Z21's wheels and suspension were separated from the main fuselage, and an exposed lightweight frame carried the front- and rear-suspension mounting points as well as the rollover hoop and the mounts for the 100-hp, four-cylinder engine, taken from the K1100 motorcycle. Though there was no windscreen or roof, BMW's wind-tunnel-honed aerodynamics promised reasonable protection from the worst of the weather. Regrettably, however, these qualities would never be put to the test as the Z21, also known as the Just 4/2, was never considered for production.

Z07, 1997

BMW designers took a calculated risk at the 1997 Tokyo Motor Show when they decided to reimagine the company's most sacred four-wheeled icon, the 507 of four decades earlier. The idea was to advance the design as if it had been subjected to the same multiple-generation development as other model lines, such as the 3-Series and 5-Series. The result was the Z07 (above), a lightweight coupe-roadster with its unashamedly 507-like retro lines set out by Danish designer Henrik Fisker, who would later become Aston Martin's design director before founding his own car company in California.

The clearly production-ready Z07 concept generated much excitement with its potent M5 V-8 engine and exquisite

interior complete with wire-spoke steering wheel. And despite protests in some quarters that the retro approach was failing to take BMW design forward, the groundswell of enthusiasm (and offers of deposits from eager customers) was so strong that BMW tooled up a special production line for its aluminum structure. Within little more than a year it was on sale, with precious few visual changes (save for the loss of the fastback hardtop), as the Z8.

Z9, 1999 and 2000

As a statement of intent, the Z9 Gran Turismo concept of 1999 could hardly have been more emphatic. On the cusp of the new millennium and with new chief designer Chris Bangle in charge of the studio, the design showed a dramatically different way forward, jettisoning many BMW design cues previously regarded as inviolable and adding a series of new features that met with bafflement or even outright disbelief. The exaggerated bustle on the tail aroused particular hostility, and few succeeded in understanding what Bangle called the Intuitive Interaction Concept—the polished metal knob in the cabin that is nowadays much more familiar as the iDrive controller.

Whilst the Gran Turismo, with its long gullwing doors, previewed a luxury 2+2 seater sports coupe, the following year's Z9 Cabrio showed an open configuration as well as a simplified version of the interior control system. Both cars remained intensely controversial for a number of years, especially among enthusiasts who believed luxury cars should be conventionally beautiful rather than confrontational. However, the Z9 paved the way for the even bigger shock of the E65 7-Series of 2001, and now, from the perspective of 2016, it is hard to understand what all the fuss was about.

X Coupe, 2001

History has been rather kinder to other BMW concept models than to the X Coupe of 2001. To most observers at the time, the design was the most shocking thing yet seen from chief designer Bangle; he had already been pilloried in the mainstream press for the Z9 concepts, and this most recent arrival was seen by many media as proof that he was on an altogether different planet.

The truth was that few were able to understand the design, let alone its intended function or the reasons for its many startling features, among which was an asymmetric body whose trunk hinged up from the rear to allow access to one of the rear seats. The X Coupe's scale was confusing too: without anything to measure it against, it had the allure of a neat, compact sports car, yet its actual dimensions were macho and huge, with massive wheels, a high roof, and substantial ground clearance. Nowadays we recognize those as the traits of what BMW calls a "sports activity coupe" along the lines of its own 2008 X6—and Bangle's polarizing design must take some of the credit for opening up this new and now highly fashionable niche of the market.

Mille Miglia, 2006

Though never displayed at any major international motor shows, the Mille Miglia concept coupe was the first of BMW's modern homages to the heroes of its back catalogue. Resisting the temptation to slavishly copy the classical icon (as happened

with the Z07 and its 507 template), the Mille Miglia is a genuine reimagining of the special Touring-bodied lightweight 328 coupes that dominated the 1940 running of the Italian road race.

The cues are all there, but reshaped into a modern context: the tall and slender grille, extending back into the hood; the large, round headlights set inward from the fender line; and the divided windshield with the twin wipers mounted at the top. The drilled disc wheels pay tribute to the originals, and the vertical slashes behind the rear-wheel arches echo the louvers in the 1940 car's hood sides. A particular delight is the long, tapering aerodynamic tail with its asymmetrically truncated rear window and faired-in rear wheels.

CS, 2007

Reviving a familiar and very emotive BMW designation, the 2007 CS concept was widely tipped by BMW management as a future 8-Series—the luxury four-door coupe, a format fashionable at the time, which would form the pinnacle of the company's premium passenger car lineup. As a flagship luxury car, the CS was imposing—not for its expected bulk or stature but for its low, sleek, and extremely wide stance and for its new interpretation of the BMW double-kidney grilles as larger and more aggressively thrust forward. The bodysides introduced sharp swage lines that would be seen on later production models, and the fastback rear style influenced subsequent coupe designs, marking the beginning of the break from the intricate surfacing of many Bangle-era models.

In terms of interior design, the CS was especially finely crafted and even included round ceramic control knobs on its metallic bronze dashboard. Expectations for the production car were high, but the project was canceled in the wake of the global financial crisis that kicked off at the end of the decade.

Gina, 2008

Reflecting his desire to explore the feeling of human interaction with the vehicle, Chris Bangle's Gina project (below left) was more of an off-piste research venture than a finished concept car. Its most headline-grabbing innovation was the use of a soft fabric skin stretched tight over a structural skeleton, in this case a carbon-fiber frame laced with wire and other materials. Bangle's big idea was to present a seamless but also flexible surface, allowing the doors to open without the need for an unsightly panel gap at the front—instead, the surface material crinkled at the point of the door hinge. The engine, likewise, was accessed by unzipping the fabric along the centerline of the hood. The design, said Bangle, was the first step on the way to developing materials that would allow future vehicles to change their shape according to the driver's whim and to suit the roles they were being applied to.

M1 Hommage, 2008

It is a high-risk business reinterpreting an acknowledged design classic to suit a more modern era, and it is even more fraught with danger if the classic in question is not only a high-profile supercar but also a design penned by a true maestro such as Giorgio Giugiaro. So as the thirtieth birthday of the 1978 M1 approached, BMW's design team, under Adrian van Hooydonk, must have questioned the wisdom of producing a tribute to the masterpiece, especially as two classics of similar time frame and stature—the Ford GT40 and Lamborghini Miura—had just been revisited by 1990s studios to a generally lukewarm reaction.

Yet BMW need not have worried. The M1 Hommage (opposite, top), true to its name, proved to be an exciting evocation of the 1978 design, capturing the spirit and the excitement of the original, although the proportions had been dramatically exaggerated and the design language had swapped Giugiaro's flat planes and razor edges for contemporary sweeping curves and a much more complex blending of surfaces, lines, and graphics. No line, shape, or surface was the same as Giugiaro's—the proportions were distorted, and the graphic elements parodied rather than copied those of the original. But the result was universally applauded as a modern supercar that was instantly recognizable as 100 percent M1. The only real disappointment was that the Hommage was only ever intended as a study—it had no engine or transmission, no interior, and no plan for production.

Vision EfficientDynamics, 2009

For BMW, as a forward-looking company, the radical Vision EfficientDynamics was not just a brilliant showcase for advanced engineering but—perhaps even more vitally—a decisive move in terms of imaginative visual design.

Presented in 2009 amid the gloom of the financial crisis, the 2+2 coupe concept's dramatically low proportions, complex multilayer surface language, and electric-blue highlights interweaving with the pure-white and piano-black bodywork made it seem like a vision of the future made incarnate. And indeed it was, for concealed beneath the intricately layered exterior lay further layers of futuristic technology—a plug-in hybrid powertrain that promised performance on par with an M3 combined with the economy of a microcar.

The excitement was palpable and the reaction instantaneous: this was unquestionably a quantum leap on almost every level of vehicle design, marking it out as perhaps the most important concept car in BMW's history so far, as well as one of the most significant in the whole industry. Spurred on by the intensity of this response, BMW set about turning the Vision EfficientDynamics into a series-production model and, true to the promise of the concept, the i8 launched in 2014 as the first supercar of the modern environmental age—as green as an electric car for the everyday commute, but fast and fun as soon as the desire arose.

Vision ConnectedDrive, 2011

Outwardly, the Vision ConnectedDrive (below) was just another compact, fun-loving two-seater roadster; it could have been developed into a more provocative follow-up to the classical Z4 or perhaps scaled down to make a Mazda MX-5 competitor. But that was not the role it was designed for. Rather than delivering the conventional on-the-road thrills of a regular sports car, this concept sought to heighten the driving experience, and also to make driving easier and more convenient, by networking the car to its surroundings.

Claiming in a press release at the time of its 2011 presentation that the concept was "the most intelligently connected vehicle of all time," BMW packed the two-seater with everything it knew about communications, connectivity, networking, and information gathering. The big idea of connected driving saw the vehicle drawing in information from the road and its surroundings through color-coded

fiber-optic channels, and bringing this information to the driver at precisely the right time. But if that sounds like motor-show gloss, even a tiny sample of the concept's innovative features shows how perceptive it really was: it launched a head-up display, automatic smartphone integration, gesture control, smart mapping, self-parking, automatic emergency braking, and integrated displays.

328 Hommage, 2011

Broadly similar in its open two-seater architecture to the Vision ConnectedDrive presented just two months earlier, the 328 Hommage (above and below) shifted the focus—as its title would suggest—away from communications and networking toward a much purer and more stripped-down sports-car appeal.

Referencing the iconic sports roadster from the 1930s, the Hommage has a low driving position, cutdown doors, and a taller and narrower interpretation of the double-kidney grille.

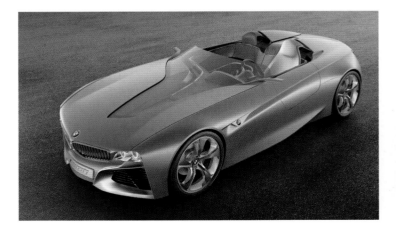

The frameless windshield is low and steeply raked, and twin fairings flow rearward from the bucket seats along the rear deck. In its configuration and its stance, the 328 Hommage is entirely modern, especially at the rear; it is through the respectful adaptation of the original's distinctive details as graphic design elements that the modern car distinguishes itself. Most striking are the drilled alloy wheels with their twin-eared knock-off hubcaps, the twin diagonal leather hood straps on each flank and on the top of the hood, and the taped-over circular headlights.

Pininfarina Gran Lusso Coupe and BMW Vision Future Luxury, 2013–2014

These two large luxury cars were designed by Pininfarina and the BMW studio at the same time, with what BMW describes as a creative exchange of high-end luxury thinking. At the same time, too, BMW's studio was completing its work on the 2015 7-Series, making the interplay between the three prestige models an intriguing issue. Pininfarina is noted for the classy elegance of its designs, and its two-door Gran Lusso coupe (pictured top and above right) retains the BMW-typical long wheelbase and hood, with the greenhouse set well back; the BMW grille is angled slightly forward but rolls upwards fractionally into the hood. The shoulder line is smooth and high, and the tail—with its very slender full-width taillights—is comparatively short.

With four doors rather than two, BMW's in-house Vision Future Luxury (pictured next page) has a weightier feel, and the narrower laser lights and extended grill roll-round make the hood appear longer. The side surfaces receive complex hand sculpting, which BMW says is beyond the capabilities of even its best computers, and the high-set tail carries a shimmering strip of OLED lights that reinterpret the familiar BMW L shape. All three designs include an air breather just behind each front-wheel arch, the vent tapering rearwards to bring depth to the sill area.

Both concept interiors celebrate elegant materials, such as forty-eight-thousand-year-old kauri wood and the finest leathers and wools. But while BMW's design centers on head-up displays, multiple touchscreens, and gesture control, Pininfarina's treatment appears closer to that of a feasible production car.

3.0 CSL Hommage, 2015

The third and perhaps most emotive of BMW's Hommage series of back-catalogue tributes was 2015's CSL (opposite page), unveiled—as were the previous two—at the Concorso d'Eleganza Villa d'Este in the Italian lakes region.

The elegant 3-liter CS coupes were BMW's halo cars from the late 1960s onward and grew steadily in power and general wildness in parallel with the increasing success of competition versions on Europe's racetracks. The series reached its high point with the wild 3.0 CSL, complete with an aggressive rear spoiler that ensured it instant fame and the Batmobile nickname.

BMW's 2015 invocation is even longer and wider, at 5m and 2m, and, in its lurid yellow shade, even wilder than the original. In profile, the design has the same proportions and cabin shape as the original, but in three dimensions the bodywork soon reveals itself as vastly more complicated: around the nose and front-wheel arches, BMW has applied its latest layered design theme, with the arches wrapping over onto the front hood—complete with the black longitudinal splitters so characteristic of the Batmobile. At the rear the design is more complex still, with the layered technique producing a fascinating swirl of light as the rear lights swing inwards, then back out again to travel up the huge spoiler and across to the other side of the car. The intricate path taken by the LEDs makes for one of the most striking light signatures ever seen. Inside, the cockpit is competition-style high tech, but not to the exclusion of period detail—a stylish wooden panel extends the full width of the dashboard just below the windshield, just as it did on the 1968 car.

9

THE VON KUENHEIM YEARS, PART I

Bound for Glory

A s the fresh-faced former troubleshooter from the Quandts' ailing IWK industrial equipment offshoot settled into the BMW managing director's chair in January 1970, there must have been many who questioned the wisdom of Herbert Quandt's choice of chief executive officer. At forty-two, Eberhard von Kuenheim was younger than all the other board members (something that would be repeated when his anointed successor, the forty-five-year-old Bernd Pischetsrieder, took over the controls almost a quarter century later); worse, though the newcomer was a trained engineer and had succeeded in "bringing the spinning aeroplane of IWK back under control," he palpably lacked automotive experience.

Yet any doubts there might have been were soon dispelled as von Kuenheim's initially cautious approach began to evolve. The company was already in good shape, having emerged from the 1967–1968 mini-recession with a sales jump of 23 percent and a turnover busting the DM 1 billion barrier for the first time. Paul Hahnemann's sales and dealer development program was in full swing, and he had even contemplated taking over the struggling Lancia company— only for Fiat to snatch it the following year at a lower price. The new Dingolfing factory and the promising 5-Series, successor to the model that had saved BMW from ruin in 1961, were ready to roll, the economy was booming again, and things were looking good.

∧ **Herbert Quandt (center) and his wife are taken on a tour of the new Dingolfing plant, then building the new 5-Series.**

‹ **The first 5-Series had the important task of following up the car that had saved BMW.**

BMW 5-Series (E12), 1972–1981

Replacing the successful Neue Klasse models, which had saved BMW in 1961, the E12 5-Series revived BMW's prewar numbering system. Though larger and less overtly sporty than the Neue Klasse, the 5-Series models were a big hit and set new standards for interior design and exterior style: they were the springboard for all of BMW's future growth and launched the template for a core three-range lineup that lasted into the twenty-first century.

1972: 520 and fuel-injection 520i launched at Frankfurt show, with 115 and 125 hp respectively.

1973: Six-cylinder 525 added, with 145 hp.

1974: 1.8-liter 518 launched in response to fuel crisis.

1975: 528 released in Europe and 530i in North America.

1977: 520 available with six-cylinder M20 engine (122 hp); 528 gains fuel injection; Motorsport-modified M535i with five-speed gearbox and 218 hp announced.

1981: Replaced by E28 model after 697,400 units built.

∧ **In charge: Eberhard von Kuenheim (left) recruited managers from outside BMW, including Bob Lutz (right).**

Nevertheless, von Kuenheim was looking further ahead—further even than the next six years, during which production numbers would double. This truly long-term approach kept him in the driving seat for two and a half decades: by 1993 car production would quadruple, motorcycle sales triple, and sales revenue increase eighteen-fold—and all the while there were operating profits in every single year. Such results are exceptional in any large industrial company, let alone one requiring such heavy investment and subject to such strict legislative control as an automaker. Von Kuenheim steered through almost one hundred new cars and thirty motorcycle models during his time as CEO, transforming BMW from a medium-sized regional manufacturer to an influential player on a global scale—and overtaking its traditional rival, Mercedes-Benz, in the process.

For all the outward signs of success, however, it was not all plain sailing. Reflecting his belief that having the right people is the key to an effective operation, von Kuenheim began by making changes in the boardroom. His most controversial early move was to release Hahnemann, the extrovert sales manager who had done so much to build up BMW's presence in the field; his aggressive style of salesmanship was no longer appropriate, argued the new chief, but such was Hahnemann's popularity among the workforce that his ouster prompted a symbolic strike—one of the few in BMW's history. Importantly, von Kuenheim brought in talent from outside, including managers from Ford and Porsche as well as Swiss-American Bob Lutz from GM Opel; Lutz would soon set up BMW Motorsport GmbH and in turn lure Jochen Neerpasch from Ford.

Engineering talent was in high demand in the early 1970s as emission controls were tightening in the vital US market, forcing European automakers to develop parallel ranges of "de-smogged" engines, often with larger capacity, higher gearing, and frequently tricky add-on technologies. A big and very public technology push came in summer 1972 when the Olympic Games took place in BMW's home city of Munich and the company enjoyed its highest public profile ever. Its dramatic headquarters—a "four-cylinder" high-rise building—was in place, electrically powered BMW 1602 models led the athletes' parades, and Paul Bracq's stunning Turbo concept wowed every onlooker. As a fascinating sign

of the times, this racy 200-hp supercar, which would later influence the M1, was billed as a safety research vehicle. The subtle message here was that while other carmakers were presenting bulky, intimidating, and downright ugly safety research prototypes, BMW could do the same thing with style and sporting flair.

Industrially, however, the biggest event for BMW was the inauguration of the Dingolfing plant and the launch of the 5-Series (E12). Eagerly awaited as the replacement for the Neue Klasse sedans that had saved the company, the 5-Series returned to the prewar system of capacity and size-based model designations and also marked a subtle shift in market positioning: the lines

∧ **Motorcycle sales picked up again following the move of production to Berlin and the launch of the new /5 generation, with up-to-date features such as electric start.**

The Munich Olympics in 1972 gave BMW the chance to highlight its inventive capacity with the futuristic Turbo concept and electric versions of the 1602, which led the athletes' parade.

penned by Bracq were fresh, modern, and classy, and the interior marked a major advancement in ergonomics and clarity of information. There was a big shift in the way the cars drove, too: the engines were sharper and smoother, and the chassis was given a less overtly sporting tune, to the benefit of comfort and refinement.

For some, this was seen as a betrayal of BMW's sporting heritage, going soft on what made a BMW a BMW. But the formula was keenly judged, and the initial 520 and 520i were instant successes with some sixty thousand buyers by the end of their first full year on sale; other versions soon followed, including the six-cylinder 525i and, in a brilliantly quick response to the first oil crisis in late 1973, the 518 economy version. The oil shock, the doubling of fuel prices, the queues at filling stations, and the Sunday driving bans brought the first real anticar sentiment to Europe, but BMW kept its nerve and kept production rolling while others laid off workers and shuttered plants; the 518 proved to be the biggest seller of a range and would last into 1981, with almost seven hundred thousand of this generation sold.

In deference to this crisis and the subsequent recession, BMW had withdrawn its potent 2002 Turbo, adding the "economy" 1502—complete with a low-compression engine— at the other end of the scale. This, too, was well judged for the times, showing that BMW quality, comfort, and style could be convincing selling points even when not accompanied by a powerful engine. Von Kuenheim, for his part, drew the line at the 1502, stating that anything below that price point would not show the refinement necessary for a BMW—again demonstrating how far BMW had come since its bubble-car days barely a decade earlier.

By summer 1975 BMW had to lay on extra production shifts to cope with demand; by now, too, the 5-Series had made the '02 models look somewhat dated, and the brand-new 3-Series (E21), launched that June, gave a further lift to the sales curve. Reinterpreting the themes of the 5-Series in a smaller, more coupe-like two-door format, Bracq again scored a bull's-eye with the new 3-Series: its wedge profile gave it a keener, sportier look; its compact dimensions and sharp rack-and-pinion steering made for excellent agility; and its driver-focused, cockpit-style dashboard marked a new industry high in interior design.

BMW 3-Series (E21), 1975–1983

Larger, more modern, and more comfortable than the '02 models it replaced, the 3-Series brought BMW to a broader and less sports-focused audience. BMW's first million-unit seller, the compact two-door was agile and entertaining to drive, and its driver-oriented cockpit set an interior-design trend for the industry.

1975: 316, 318, 320, and 320i launched, the latter two featuring four headlights.
1977: Six-cylinder 320/6 with new M20 engine giving 122 hp replaces 320; Baur cabrio version added.
1978: Sporting 2.3-liter 323i six with 143 hp becomes an instant success.
1981: Entry-level 315 model (75 hp) added; 318 gains fuel injection.
1982: Principal range replaced by new E30 models, but base 315 continues.
1983: 315 ceases production, bringing overall total to 1,364,039 units.

Sales really took off with the introduction of the smaller 3-Series, which replaced the '02 models. The interior set new standards of style and clarity.

As with the 5-Series, sections of the press complained that the sporting focus of the '02 models had been diluted and that the 3-Series had gone soft. But the sales figures spoke for themselves: the 3-Series sold over 130,000 units in its first year and edged BMW's overall production ever closer to the steady 350,000 being built by Mercedes-Benz each year. Behind the scenes, however, there were personnel changes, with both Lutz and Bracq leaving and Claus Luthe, the man behind the revolutionary NSU Ro80, arriving as head of design.

While some commentators accused BMW of "freewheeling spending" on research and development, capital equipment, and recruitment, the company countered that it was tooling up for the future, beginning work on diesel engines, reactivating the large-car programs that were slowed down during the recession, and prioritizing the development of a brand-new small-block six to fit into the 3- and the 5-Series. Available with the new M20 unit from late 1977, the turbine-smooth 320/6 and 520/6 instantly raised the bar for sophistication, while the hotshot 323i, with its free-revving 143 hp, surpassed the fondly remembered 2002 tii for keen-edged sportiness and hard-to-resist desirability.

While the 323i quickly became the most fashionable quality car to own, the basic 316 continued as the top or number-two seller, confirming that BMW's appeal was as much about image and quality as outright performance. This appeal propelled ever-greater production numbers each year and made the 3-Series BMW's first million-unit seller by the time the second-generation E30 3-Series model took over in 1982.

The resilient success of the BMW recipe did not escape the notice of other carmakers, jealous of BMW's ability to often increase its sales in recessionary periods when rivals were in retreat. Von Kuenheim prized BMW's carefully thought-out position, and he expressed more determination than ever to keep prices at premium levels and never to produce too many cars; at one point he reportedly even considered dropping four-cylinder engines altogether, something opposed by his respected but conservative head of vehicle development, Karlheinz Radermacher.

Pre-dating the small sixes were two distinctly high-class model lines—the 6-Series coupe, based on 5-Series engineering, and the luxury-class 7-Series that completed von Kuenheim's model palette. The style was Bracq at his best, especially on the long and elegant 6-Series (E24) with its crisp, clear surfacing and balanced proportions; the taller 7-Series sedan (E23), by comparison, came across as bulkier and more imposing, though it looked much fresher than the Mercedes SE of the day.

At the launch of the 7-Series, BMW declared itself "supremely confident, without being arrogant" at the model's prospects, signaling its return to the luxury market and a much more direct challenge to Mercedes. Ever since the mid-1960s there had been talk of a still larger, "grosser BMW" to take the company to the very highest echelons of the market; both V-8 and V-12 powerplants had been considered, the twelve-cylinder winning out over the eight but finding itself shelved once the

∧ **Designed at the same time as the 6-Series, the 7-Series gave BMW a strong offering in the luxury class dominated by Mercedes.**

❯ **The distinguished 6-Series coupe was one of Paul Bracq's finest designs. Pictured is a later 635i version.**

BMW 6-Series (E24), 1976–1989

The elegant 6-Series coupe was based on the architecture of the 5-Series sedan but with larger, fuel-injection six-cylinder engines and additional luxury equipment. Replacing the famous 3.0 CSi coupe series that was so successful in motorsport, the M635 CSi was the first of the M cars. The design remained in production for thirteen years.

1976: 630 CS (185 hp, carburetor) and 633 CSi (fuel injection, 200 hp) announced.
1978: 635 CSi with 218 hp becomes fastest four-seater on the market at 220-plus km/h.
1979: 628 CSi with 184 hp replaces 630 CS.
1982: 633 CSi dropped from range.
1984: Motorsport-developed M635 CSi launched, with 286-hp, 3.5-liter, twenty-four-valve engine from M1 supercar. Price is nearly double that of other models.
1987: 628 CSi dropped, leaving only M635 CSi in production.
1989: Production of M635 CSi ceases.

^ Turbocharging provided a quick route to high power in the 745i, but the technology was immature and the car problematic.

second energy crisis hit home in 1979. Yet, still feeling the urge to beat Mercedes in the power stakes, von Kuenheim decided instead to equip the 7-Series with a turbocharged version of the 3.3-liter straight six, giving 252 hp, to claim the title of the segment's power champion. It was a move that BMW was to regret, for the 745i proved to be temperamental to drive, heavy on fuel, and, worst of all, unreliable in service. At a stroke, much of the precious goodwill BMW had built up toward its luxury models was lost, and it was only after two further engine and transmission iterations that the company arrived at a halfway satisfactory solution. But by this stage the damage was done, and BMW would not return to turbocharged gasoline engines until after the start of the new millennium.

At the 1978 shareholders' meeting, von Kuenheim told investors that 30 percent of overall engineering time was devoted to meeting the many international regulations, and at a technical conference later that year, BMW demonstrated three experimental powerplants to reporters: a high-compression 2.4-liter gasoline six; a similar engine able to run on either three or all six of its cylinders, depending on load; and a six-cylinder 2.4-liter turbocharged diesel giving 125 hp.

It was the latter that proved most impressive, and, convinced that diesel could combine both economy and responsiveness, BMW teamed up with Steyr in Austria to

▲ **The sweet and smooth-running M20 small six was specially designed for the 3- and 5-Series and made them uniquely sophisticated in their sectors. The 323i was a hotshot favorite.**

pursue diesel development and manufacture. Unfortunately, the results were disappointing and BMW had to take full control; the eventual production engine would only be ready for European applications in the 1984 model year, and the 524td—then the fastest diesel in the world—was sold for the 1985 and 1986 model years in the United States.

As the 1970s rolled into the 1980s, BMW became a major operation: von Kuenheim planned to raise annual output to 350,000 units by 1982 and launched the construction of an integrated research and innovation center—called the FIZ—centrally positioned so that its two-thousand-plus engineers could enjoy direct links with the manufacturing plants as well as the corporate headquarters.

More of the Same

On two wheels, 1981 saw the first of four victories for the BMW R80 G/S enduro motorcycle in the grueling Paris–Dakar rally, while the debut of the second-generation 5-Series (E28) in the summer was eagerly anticipated by an industry impatient to see what BMW's new direction would be. Yet the vehicle unveiled surprised everybody, for it looked like no more than a mild facelift of the existing car, itself nine years old. The sense of disappointment was clear: a thrusting and energetic young company known for its innovative and

BMW 7-Series (E23), 1977–1986

The first-generation 7-Series was BMW's first competitor on par with Mercedes-Benz's luxury SE flagship. Its engineering was derived from the outgoing 2500/2800 big sixes, while its crisp four-door body shared themes with the 6-Series coupe. The 7-Series introduced a host of technological firsts, including fully electronic management of ignition and fuel injection. For their day, these models were eager performers and adept handlers, reflecting the successful qualities of the smaller 3- and 5-Series.

1977: Launched as 728 (170 hp), 730 (184 hp), and 733i (197 hp).
1979: 3.2-liter 732i, with 197 hp, is world's first car with Motronic fully electronic mapping of fuel injection and ignition; 735i (218 hp) replaces 733i.
1980: Turbocharged 745i with 252 hp becomes range's flagship, as well as Germany's most powerful luxury sedan; three-speed automatic and ABS standard.
1982: Facelift brings wider and squarer BMW kidneys set into main grille.
1983: 745i engine recalibrated and four-speed automatic fitted for improved drivability.
1984: 745i engine capacity moves to 3,430 cc; ABS now standard on all models.
1986: Last of the 285,000 E23 7-Series models come off the line.

The second-generation 5-Series (above) looked almost identical to the outgoing model (below), but there were big changes under the skin.

exciting designs had come up with such a cautious follow-up, as if unsure of its own direction.

Yet BMW's assurances that it was indeed all new were borne out by statistics such as a 100kg reduction in weight thanks to the use of computer techniques in its structural design, a dramatic improvement in aerodynamics, and an all-injection engine lineup with the exception of the most basic 518. On the road the second-generation car shone: smoother, more comfortable, and more cultivated, with a host of new technologies, it represented engineering advance rather than stylistic enhancement, something BMW customers would soon become accustomed to. The dull design did not worry the buyers, who bought more than seven hundred thousand examples over the model's seven-year lifespan, particular standout versions being the first M5 super sports sedan and the short-lived 525e and 528e, with their low-revving engines giving optimum efficiency and refinement. With less friction, vibration, and stresses, this engine was ahead of its time: only now, in the second decade of the twenty-first century, are automakers beginning to recognize the value of downspeeding.

The launch of the second-generation 3-Series (E30) in 1983 fired the starting gun for a further phase of technical innovation and expansion; already, subtle signs of the influence of chief designer Claus Luthe were beginning to show. Superficially the 3-Series was no less evolutionary than its 5-Series counterpart—it was convincingly fresh, neat, and modern in its look—and it was just as warmly received. The range soon expanded to include convertibles, station wagons (under the revived Touring label), four-wheel drives, diesels, and the razor-sharp M3 near-racer, which went on to dominate the European motorsport scene and set new standards for road-going thrills.

By the end of 1983, BMW was flushed with success, having scooped the Formula One World Championship with its turbo engine in the back of Nelson Piquet's needle-nosed Brabham BT52, and with road-car sales leaping by a double-digit percentage. A third factory—Regensburg—was planned for 1987, but von Kuenheim admitted to the *Financial Times* that "the time of big increases in production and profit is at an end." The *FT* observed that nearly every person in West Germany who could afford a new car already had one, leaving only replacement to provide demand, and that the West German

BMW 5-Series (E28), 1981–1987

The second-generation 5-Series was outwardly very similar to the original but more aerodynamic and 100kg lighter thanks to the use of computer-aided design. Innovations included Motronic electronic engine management, energy meter, service interval indicator, and onboard computer and ABS; the sportier 528i offered the choice of overdrive or close-ratio five-speed gearboxes.

1981: Launch lineup includes 518, 518i, 520i (125 hp), 525i (150 hp), and 528i (184 hp).

1983: 524td diesel (115 hp) for Europe; 533i is fastest sedan in the United States; low-revving 525e (Europe) and 528e (US) economy models.

1984: Sporting M535i with 218 hp, lowered suspension and aerodynamic body kit.

1985: High-performance M5 replaces M535i, with Motorsport-developed twenty-four-valve 286-hp engine from M635 CSI and 245km/h top speed.

1987: Replaced by new-generation E34 after 722,348 examples built.

BMW 3-Series (E30), 1982–1993

The second-generation 3-Series built upon the success of the first, adding wagon, four-door, and convertible versions and selling more than two million units in eleven years' production. As with the 5-Series, equipment now included an energy meter, check control, and an onboard computer, and the upgraded chassis had rear suspension geometry improved to reduce liftoff oversteer.

1982: Model launched as two-door 316, 318i, 320i, and 323i (139 hp).
1983: Four-door body style added.
1985: 325i with 171 hp replaces 323i; 324d diesel, with 86 hp, and four-wheel-drive 325iX added.
1986: Full cabriolet 325i and high-performance M3 added. Developed by BMW Motorsport, the M3 had bodywork changes and a 2.3-liter, sixteen-valve, four-cylinder DOHC engine giving 195 hp.
1987: Exterior facelift; 318i gains new engine with Motronic control; 324td turbodiesel with digital motor electronics and 115 hp; Touring (four-door wagon) models announced.
1988: Touring enters production; M3 Cabriolet added; M3 Evolution has 215 hp.
1989: Two-door 318iS with new Motronic-managed sixteen-valve engine giving 136 hp.
1990: M3 Sport Evolution brings power to 238 hp and top speed to 238km/h.
1990: New-generation E36 sedan revealed for a 1991 production start; sedan production ceases but Cabrio and Touring continue.
1993: Cabriolet production ceases.
1994: Touring production ceases.

Hubert Auriol leaps to victory on the 1981 Paris–Dakar rally raid, the first of four wins for the R80 G/S. The model would go on to be a major commercial success in the following decades.

population was declining. Additionally, Germany's success had triggered a strong rise in the value of the D-Mark against the US dollar, making American exports less profitable; some commentators suggested BMW should diversify into other industries, just as Mercedes had just done with its acquisition of aircraft maker Messerschmitt-Bölkow-Blohm (MBB).

Von Kuenheim countered by saying that "any company we buy must be big enough to warrant our management time but small enough not to damage BMW should anything go badly wrong." These would prove to be prophetic words both for Mercedes-Benz and MBB and, a decade later, for BMW's fraught relationship with its big acquisition, Rover.

In an interview with the *FT* at the 1986 Frankfurt show, von Kuenheim confirmed he would seek to expand volume by

▲ **A four-door E30 poses outside BMW's "four-cylinder" headquarters and museum.**

only about twenty thousand vehicles each year and that if BMW were to produce some six hundred thousand cars a year it would be in danger of losing the individuality that made its models so attractive. His quiet, step-by-step confidence was based not only on the elegant second-generation 7-Series, which he knew heralded a new era of highly stylish BMWs, but also on his well-kept secret that BMW was about to deal a body blow to

its Stuttgart rival's pride by launching a prestigious V-12 engine and knocking the Mercedes S-Class off its perch as Germany's favorite luxury car. And what is more, at that point no one at all had the slightest inkling of an even bigger event that would turbocharge every automaker's business—the fall of the Berlin Wall, the domino-style collapse of the Soviet bloc, and the unfolding of many more lucrative markets around the world.

99

10
Power, Performance, and Art

I n the mid-1970s, French racing driver, auctioneer, and car enthusiast Hervé Poulain called up his friend Jochen Neerpasch, who just happened to be BMW's director of motorsports at the time, with an outlandish idea. Poulain, a self-confessed lover of both cars and art, said that he wanted to compete in the 24 Hours of Le Mans and turn racing cars into art and that BMW would be the ideal partner. It helped that Poulain also counted American artist Alexander Calder among his circle of friends, and it was not long before the first BMW Art Car—in Calder's way-out primary colors—was completed and lining up for the start of the French classic. What none of these men could have known at the time was that theirs was the first in what would turn into a long line of BMW Art Cars, a series of dynamic canvases that would span the next four decades and bring in artists of worldwide repute such as Andy Warhol, David Hockney, and Jeff Koons as well as introduce lesser-known artists from countries including South Africa, China, Iceland, and Australia to a global audience.

In 2015 BMW hosted a reunion of almost all of the "rolling sculpture" artworks in Munich, prior to a world tour. The latest twist in the Art Car story has been the announcement that the eighteenth BMW Art Car will be the first in the series to represent a collaboration between two artists. Beijing-born Cao Fei, who works with video and digital media, and John Baldessari, an American conceptual artist some fifty years her senior, will work together to add the new BMW M6 GT3 race car to the collection, and the results of their endeavors will be seen in 2017.

Alexander Calder, 3.0 CSL, 1975

New York sculptor Alexander Calder, best known for his colorful and gently moving mobiles, gave the Art Car program a dramatic start with his bold 3.0 CSL race car, bedecked in vivid primary colors that swirled asymmetrically around the whole Batmobile-style body. The 480-hp car, with race number 93, was driven by Sam Posey, Jean Guichet, and Hervé Poulain in the 1975 24 Hours of Le Mans but was forced to retire after seven hours with a broken driveshaft. The BMW was one of Calder's last works, as he died the following year.

Frank Stella, 3.0 CSL, 1976

Noted American painter and printmaker Frank Stella was next to take up the Art Car baton, clothing another 3.0 CSL race car with a geometric black-and-white grid design, much like giant graph paper. Cutout lines symbolize the mechanical elements beneath the bodywork and, like Calder's high-speed mobile the year before, the Stella CSL was a firm favorite with the crowds at Le Mans.

Roy Lichtenstein, 320i, 1977

Roy Lichtenstein shot to fame in the 1960s as one of the leading figures in the new Pop Art movement, and his 1977 Group 5 BMW 320i racer (right) was the perfect vehicle to display his distinctive comic-strip technique, with its giant spotted halftone patterns and wavy primary-colored contours. "The painted lines symbolize the road the car has to follow and the artwork also portrays the surroundings through which the car is being driven," commented the artist.

Andy Warhol, M1, 1979

As one of the biggest names in the art world to create a BMW Art Car, Andy Warhol approached the M1—another race car—with a professional touch. The blurry but colorful hand-brush finish (above, foreground) was said by Warhol to portray a sense of speed—very appropriate for this 470-hp racer, which finished second in its class in the 1979 24 Hours of Le Mans.

Ernst Fuchs, 635 CSi, 1982

Austrian professor of art Ernst Fuchs, cofounder of the Vienna School of Fantastic Realism, was the first non-American invited to create a BMW Art Car. This was also the first time a standard production car was decorated in this way, and Fuchs's fine-art illustrative technique was deployed to dramatic effect on the dark blue-black coupe, with spectacular flames emerging from the wheels, spreading along the hood, and engulfing the roof.

Robert Rauschenberg, 635 CSi, 1986

Robert Rauschenberg, another influential figure on the 1960s New York Pop Art scene, was a painter and graphic artist who worked with cut-up techniques to expand the boundaries of art. His BMW 635 CSi was the first Art Car to use photographic processes: on the right-hand side is French painter Jean-Auguste-Dominique Ingres's enigmatic reclining nude *Grand Odalisque*, while the driver's door carries a work by Italian artist Bronzino.

Michael Jagamara Nelson, M3 Group A, 1989

Using a black BMW M3 sedan as his canvas (above), Australian artist Michael Jagamara Nelson spent seven days applying by hand the ancient, indigenous art techniques he learned from his grandfather. The intricate, interwoven, mosaic-like designs symbolize landscapes and animals, said the artist.

Ken Done, M3 Group A, 1989

Fellow Australian artist and designer Ken Done took a highly colorful and graphic approach to his creation, like Nelson on the basis of a BMW M3 race car. The bodywork, signed in large script on the rocker panel, conveys a powerful sense of vibrancy and movement from the front—hot and sunny—to the rear, where dark colors predominate and the stars are coming out. Hidden in the abstract portrayals are parrots and parrot fish, says Done.

Matazo Kayama, 535i, 1990

Snow, the moon, and cherry blossoms were the themes running across the bodywork of Matazo Kayama's BMW 535i sedan. The Japanese painter, sculptor, and printmaker approached these subjects in a new and intricately detailed way by using airbrush processes as well as the traditional Japanese techniques of kirikane (metal cut) and arare (foil printing).

CHAPTER TEN

César Manrique, 730i, 1990

The large dimensions of a BMW 730i luxury sedan gave Spanish artist César Manrique an extensive canvas on which to express the bold and colorful sweep of his visual ideas. As a graphic designer, sculptor, architect, landscape designer, and painter, Manrique was also a conservationist, and his work on the BMW Art Car was intended to depict a harmonious relationship between nature and technology.

A. R. Penck, Z1, 1991

A. R. Penck is the alias of Dresden-born Ralf Winkler, a multitalented artist and neo-expressionist painter known for his stick figures, as well as a free jazz drummer. Choosing a bright red BMW Z1 as his surface, he adorned the compact two-seater sports car with dozens of bold, black stick people, staring eyes, lions, crocodiles, and other mystical symbols, creating echoes of ancient cave paintings in the process.

Esther Mahlangu, 525i, 1991

Esther Mahlangu was not only the first female artist to be invited to design a BMW Art Car but also the first person from the African continent. She decided to paint her BMW 525i in bright pinks, blues, and purples with the traditional geometric and ornamental shapes of her Ndebele ethnic tribal art. Every inch of her car's surface is covered—even the wheels have colorful ornamental discs to add movement to the design.

Sandro Chia, M3 GTR, 1992

Convinced that cars are some of the most coveted objects in today's society and are thus exposed to many stares, Italian artist Sandro Chia adorned his BMW 3-Series race car with layer upon layer of brightly colored and often overlapping faces that stare back at the observer. Wherever the glance falls on the low-slung racer's bodywork, there is a winking eye or a wistful smile returning the gaze, adding a new dimension of intrigue to the car's style.

David Hockney, 850 CSi, 1995

Leading international artist David Hockney took an inside-out approach to his decoration of the BMW 850 CSi Art Car (left). His aim, according to BMW, was to portray the innermost workings of the car, and Hockney's abstract componentry—realized in bright red, black, grays, and large zipper-like graphics—disconcertingly breaks up the familiar shape of the 8-Series body. Clearly visible on the left-hand side are the steering column, the driver, and a dog in the back seat.

Jenny Holzer, V-12 LMR, 1999

Jenny Holzer is an American artist who uses installations, graphics, projections, and large-scale public displays to put her messages across, the messages themselves being texts, slogans, and sometimes simple but provocative single-line statements. For her, the medium of BMW's 1999 Le Mans racer was ideal—a high-speed, high-profile white canvas with *protect me from what I want* emblazoned in large silver-foil capitals over its upper surfaces, and the subsidiary message "lack of charisma can be fatal" running the width of the rear spoiler.

Olafur Eliasson, H2R, 2007

In 2007 BMW entrusted Icelandic sculptor and installation artist Olafur Eliasson with a valuable H2R hydrogen-powered research prototype, a long and narrow projectile designed for speed and efficiency. Eliasson stripped away the sleek bodywork and replaced it with a giant white steel cocoon of spiraling ribs, then covered this with thick layers of ice (above). Illuminated from the inside by a warm orange glow, the structure becomes visible as the ice melts and drips to the ground—a perplexing metaphor for today's debate on climate change.

Jeff Koons, M3 GT2, 2010

Perhaps the most visually striking of the entire Art Car series to date is Jeff Koons's M3 GT2 race car (right), which competed in the 2010 24 Hours of Le Mans. A riotous collage of every possible contrasting color and tone, its multi-hued stripes radiate backwards from the BMW's grille, as if sculpted by the sheer force of the car's speed. At the rear, a blinding flash of white, like an impact or an explosion, warns the following drivers to keep their distance. In action, it looked even more dramatic, making an already fast car seem faster still.

POWER, PERFORMANCE, AND ART

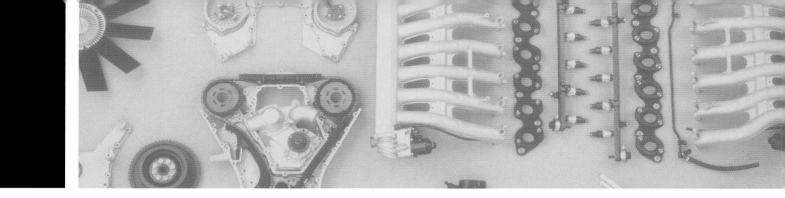

M·DM 5836

11

THE VON KUENHEIM YEARS, PART II

A New Style Emerges

In early autumn 1986, as the glamorous Mondial de l'Automobile in Paris opened its doors, BMW was everybody's hot property. So much so that suitors had lined up outside company headquarters in Munich in the vain hope of persuading the long-term majority shareholders, the Quandt family, to part with their controlling stake. America's number-three automaker, Chrysler, was rumored to have offered three times the stock-market valuation for the fast-growing company, which—for good reason—was then seen as the big prize in the auto business.

The 3-Series was selling in big numbers for a premium car—more than three hundred thousand a year—and the bigger 5-Series was holding out better than expected against the Mercedes W124, predecessor to today's E-Class; only the 7-Series, dating back nearly a decade and now looking decidedly slab-sided and quarrelsome, appeared past its sell-by date. But von Kuenheim knew he had the answer, and as he drew the wraps off the brand-new 7-Series, it was clear to every onlooker that BMW had pulled another trump card out of its pack—this time a genuine sense of style.

Apart from the obvious "wow factor" voiced by all present, Claus Luthe's design for the E32 7-Series achieved what no BMW designer for two decades had dared to do—to

break away from the upright, rectilinear template established by the 1960s Neue Klasse, which had brought such success to the company that it had come to be seen as inseparable from the brand. With admirable skill Luthe and his team had blended key signifiers, such as the quadruple round headlights, the twin-kidney grilles, and the so-called "Hofmeister kink" in the rear pillar, into a shapely body with genuine curves and contours. Most striking of all was an entirely new quality: the dynamic stance provided by the low nose, the rising waistline, the shallower glasshouse, and the high rear deck leading to the L-shaped taillights that would quickly become an additional brand hallmark.

Under the skin, the 730i and 735i—the two versions initially offered—did just what everyone expected of the company, showing the world once again how cockpit design should be done, scoring further advances in ergonomics, and introducing a whole raft of electrical and electronic systems never before seen on a car. It mattered little that in its mechanical elements the new 7-Series was a linear development of previous models; it drove very well and handled tidily and enjoyably, with excellent ride comfort thanks to its adjustable dampers—that much was expected. But, more importantly, the 7-Series achieved precisely the

BMW 7-Series (E32), 1986–1994

The second-generation 7-Series is a very important design in BMW's history, drawing level with the Mercedes S-Class in terms of engineering, equipment, technology, quality, and, for the first time, prestigious styling. Claus Luthe's lines provided a discreetly sporty allure, while six-cylinder engines gave sparkling performance and the high-level technical content included position memories for the front seats and door mirrors. But the E32 is best remembered as the platform for BMW's first V-12 engine, which enabled the company to snatch technology leadership from archrival Mercedes-Benz.

1986: Unveiled at Paris motor show as 730i (188 hp) and 735i (211 hp), both with five-speed manual or optional ZF four-speed automatic transmission.

1987: Launch of 750i with 5-liter V-12 engine giving 300 hp and with top speed regulated to 250km/h. V-12 models are distinguished externally by broader and flatter kidney grilles and are the first in the world with ultrasonic park distance control, xenon headlights, and soft-close doors and trunk. Also launched are long-wheelbase 735iL and 750iL (E32/2) versions.

1991: Six-cylinder engines replaced by advanced four-cam V-8 units giving 218 hp (730i) and 286 hp (740i). Six-cylinder 730i remains in production as a lower-cost alternative.

1994: Replaced by new E38 model, after 311,015 units produced.

impact that von Kuenheim intended, demonstrating that BMW was an innovator not just in engineering and electronics, but also when it came to style, poise, and premium image too.

Nor could the 7-Series program be seen just as a vain and costly extravagance: von Kuenheim was all too familiar with the commercial reality that while the 3-Series made up 60 percent of all BMW sales and provided 45 percent of the company's profits, the 7-Series, with just 10 percent of production volume, accounted for double that percentage in profit. Speaking to the *Financial Times*, he said that the brand renewal signaled by the new 7-Series would encourage customers to buy cars with more equipment, higher specifications, and higher prices, allowing turnover to climb more swiftly than purely numerical sales volume.

Setting BMW's lofty ambitions in stone, or rather metal, less than a year after the six-cylinder 7-Series' debut, the company presented—with barely disguised pride—the twelve-cylinder 750i version. This was the first German-built V-12 in a half century, the genuine engineering highlight that would finally cement BMW's equality in status with Mercedes-Benz. It would play an important role in helping BMW snatch technical supremacy from Mercedes, as von Kuenheim recalled in a 2004 interview:

> *Towards the middle of the 1980s, the 7-Series was doing very well. We were convinced that our six-cylinder engine was the best thing in world at that time, and every wisdom in those days said we should progress upwards to a V8, which was seen as the state-of-the-art engine for an elite car. All the market research said we should go to the V8, too— but I disappointed a lot of people and I said "no I am against this, full stop." Instead, I said we should go straight to the V12. That was a very important decision for BMW: it was absolutely against the conventional rules but it meant we that we jumped over the competitors. We launched the engine and went on to make two thirds of all 12-cylinder products in the world. This gave us such a high ranking, and ever since that time we have been on the same level of status as Mercedes.*

The best part of the journey.

Once again, the familiar "please fasten your seat belts". Once again, the squeal of tyres on the runway.

Once again, the end of a hard day's work. At last, even though it's getting late, your time is your own again. Things start well. A brisk walk to the car park, where your BMW 7 Series is waiting.

Now the passiveness of your journey so far gives way to enjoyable activity. The key in the door and the brief hum of the central locking sounds like a friendly "hello".

This time, fastening your seat belt is a welcome act. The cockpit's Check Control signals everything's ready for take-off. And it's good to know that it will go on monitoring all the car's vital functions every second of your journey.

Out onto the street. You immediately sense the 220 hp going to work.

Luckily, at this time of day, there's not much traffic about. Third gear, fourth, your BMW 735i reacts spontaneously. Its vitality demands attention, but never excessively.

The "precision–steering" suspension is finely tuned for easy, confident and dynamic handling, whatever your speed and whatever the conditions. A reassuring feeling of driving safety.

The motorway will be the fastest way home. In fifth gear you're cruising along at a healthy speed.

You lean back, enjoying the comfort. A sign of the new age: dynamism plus relaxation.

You're home. Time in the BMW 7 Series has passed quickly. A shame, really.

The ultimate driving machine

The new-generation 7-Series in 1986 brought elegance, poise, and real style to the BMW flagship for the first time. The sophistication was echoed in a sumptuous cabin with countless electronic innovations.

111

BMW seized the opportunity to launch the first German-built V-12 in fifty years. It was a pivotal moment in the company's bid to match Mercedes in technology and prestige.

While the prestige of the V-12 engine provided a tremendous psychological boost to everyone within the BMW organization, it was developments in the lower model series that provided the biggest fillip to the company's market performance. The third-generation 5-Series, code-numbered E34 and presented in autumn 1987, made up for the disappointment of the predecessor model six years earlier: again styled under Luthe, the design took the theme of the 7-Series and reinterpreted it in a sportier, more compact context, achieving an even more energetic wedge stance and severing any connection to the square-rigged generations that had gone before.

As far as the buyer was concerned, here at last was a BMW whose style spoke as eloquently as its engineering—a BMW with a powerful visual appeal to match its unquestioned technical prowess. Sharing six-cylinder engines with both the 3- and

7-Series, the new 5-Series was an immediate success, attracting more than two hundred thousand customers in each of its best years and giving its stodgy Mercedes rival a tough time. Systematic development kept it fresh, adding Touring wagon versions, the first truly satisfactory premium-performance diesel, four-wheel drives, the new V-8s from the 7-Series, twenty-four-valve versions of the sixes, and, perhaps most memorably, the remarkable M5—which, with as much as 340 hp from its high-revving six, was voluntarily restricted to a top speed of 250km/h.

Yet all the while, von Kuenheim was looking several steps ahead, not just to new models and technologies but also to the structure of BMW's core business. Profits had taken a slight dip in 1987 because of the major R&D investment in the model range renewal, the decline in the motorcycle business was causing concern, labor costs in Germany had accelerated way beyond those of other major economies, and the rising D-Mark was harming the competitiveness of German exports. All this set the board to thinking about improving the logistics of the manufacturing organization, creating a more flexible working-time relationship with the production staff and, just possibly, weighing up the feasibility of a factory outside Germany.

As the end of the decade approached, BMW announced that it would indeed set up a manufacturing plant in the United States, but bigger events in Europe would refocus the industry's attention in an easterly direction. Following the fall of the Berlin Wall in November 1989 and the collapse of the Iron Curtain, every automaker rushed east in search of cheap labor and factories at bargain prices. Not BMW's von Kuenheim, however: even when offered the company's old Eisenach plant, once the center of all its car operations, he brushed aside any sentimentality and gave a polite no; Škoda in the Czech Republic received the same response. Von Kuenheim's preference, he told the *Wall Street Journal*, was to start small and hire his own well-qualified workforce. It would not be until 2001, two chief executive officers down the line, that BMW would decide to set up a manufacturing plant in Leipzig, in the former east.

With BMW car sales having already surged past the half-million mark in 1989, the boom following German reunification came just in time to offset the downturns in many

▲ **The new 5-Series picked up on the more flowing style of the 7-Series, resulting in a fine sports sedan.**

BMW 5-Series (E34), 1987–1996

For its third generation, the 5-Series made a major step forward in design to match the progressive Claus Luthe–penned E32 7-Series launched the year before. The initial six-cylinder engines came from both the 3- and 7-Series and the chassis and suspension reflected 7-Series practice, while the 535i was offered with speed-sensing Servotronic power steering. The E34 M5 marked a high point, while later Touring variants prolonged the production run to nine years. An unqualified success in the market, the E34 was the first large BMW to sell over a million units, wresting the advantage from Mercedes.

1987: All-six-cylinder launch lineup includes 520i (129 hp), 525i (170 hp), 530i (188 hp), and 535i (211 hp) with optional Servotronic steering.
1988: Motorsport-developed M5 version, with a 3.6-liter twenty-four-valve straight six giving 315 hp; turbodiesel 525td (115 hp) made available.
1989: Four-cylinder 518i (113 hp) added.
1991: Touring (wagon) models introduced, with innovative double-hinged tailgate design; four-wheel drive 525iX added, and intercooled turbodiesel 525tds (143 hp) replaces 525td.
1992: M5 engine enlarged to 3.8 liters and 340 hp; V-8 engines for 530i (218 hp) and new 540i (286 hp), which replaces 535i; new twenty-four-valve engines in 520i (150 hp) and 525i (192 hp) introduced as running change with no visual differentiation.
1995: Sedan models replaced by new E39 5-Series; Touring versions continue in production.
1996: Touring ceases production, bringing grand total built to 1,306,755.

BMW's first multimillion seller, the third-generation 3-Series took a big step forward with its smooth, integrated design and remained in production for nine years. Available in five different body styles, it was bigger than the previous model and (except for the Compact) employed the sophisticated Z-axle rear suspension giving improved ride comfort and steering. It felt lower and sportier than its predecessor, and the initially coupe-based M3 was the first with six-cylinder, twenty-four-valve power.

1990: Launch range of four-door sedans includes 316i (100 hp), 318i (113 hp), 320i (six cylinder, twenty-four-valve, 150 hp), and 325i (192 hp).

1991: Two-door Coupe models added, initially as 325i, later as 320i and 318iS, with sixteen-valve 140-hp version of 1.8-liter M40 engine; six-cylinder diesel 325td (115 hp) added, later upgraded to 325tds with intercooler and 143 hp.

1993: Cabriolet versions enter production, initially as 318iS and 325i; M3 Coupe launched, using BMW Motorsport-developed twenty-four-valve six with 286 hp.

1994: Two-door Compact hatchback version launched as 316i, with simplified specification and with interior and suspension from previous generation model; M3 now available as Cabriolet and four-door sedan; four-cylinder 318tds diesel engine added, with 90 hp.

1995: Touring (wagon) launched in 318i, 320i, 328i, 318tds, 325tds forms; 325i sedan, Coupe and Cabriolet become 328i, with 2.8-liter six giving 193 hp; new twenty-four-valve 2.5-liter engine rebadged as 323i.

1996: Revised M3 with enlarged 3.2-liter, 320-hp engine linked to six-speed manual or SMG sequential transmission, available as Coupe and Cabriolet; 318ti Compact gets 1.9-liter sixteen-valve engine with 140 hp.

1997: Sporting 323i version of Compact introduced, with six-cylinder 170-hp engine.

1998: Sedan models replaced by next-generation E46.

1999: Coupe, Cabriolet, and Touring models phased out.

2000: Compact withdrawn, bringing E36 build total to almost three million.

114

SONNTAGMORGEN, 7 UHR 15, FORMEL M.

Der neue BMW M5 auf Testfahrt auf dem Nürburgring. 6 Zylinder, 250 kW/340 PS, 6900 U/min, von 0 auf 100 km/h in 5,9 s.

other markets worldwide; BMW, too, was gearing up, but more cautiously than its competitors. The FIZ R&D center in Munich was now fully open, working on long-term projects rather than any short-term cashing in on the boom market, and an all-new 3-Series was poised for an early 1991 volume production start.

The E36 3-Series was a critical car for BMW, taking over from the most successful model the company had ever built and providing almost half the firm's income. The temptation to play it safe and stick with an evolutionary design might have been strong, but Luthe's bold and innovative shape came as confirmation that the design renewal was taking place right across BMW's lineup.

In the case of the 3-Series, however, the public debate was rather more intense. While there was no quarrel with the neat, low-nosed, high-tailed wedge shape, or even the lowered and widened double-kidney grilles set into a sheet-metal frontal panel, it fell to the headlights to stir up the real emotion. Gone were the trademark quadruple circular units, replaced by smoothly integrated rectangular glass covers through which the shape of the four round projectors could just be made out.

▲ **Sunday morning, 7:15 a.m.—driving Formula "M." BMW promotes the new M5, with 340 hp, on the Nürburgring.**

Though some complained that the BMW "face" had been lost, it was an altogether more modern and aerodynamic approach, correcting the major failing of the outgoing model, and the protests soon faded away.

With the thoroughness that had always been expected of BMW, Luthe followed the four-door sedan with a two-door coupe: in the interest of the perfect sporty look, every single body panel had been freshly designed. This was a major success with the public, consistently selling six-figure volumes (especially after the convertible version joined the range), and formed the basis of the new grown-up M3, which forsook the youthful racetrack dynamics of the first model in favor of the searing energy of a 3-liter BMW Motorsport straight six producing more power than even the legendary M635 CSi.

A NEW STYLE EMERGES

The special V-12 engine developed by BMW for the McLaren F1 supercar found success at Le Mans (above), winning outright in 1995; it had few technical links with the 750i luxury-car engine (below) save for its V-12 layout.

The organization was relatively slow to add the Touring version, as customers had been happy to continue buying the old-shape model into 1994; by that stage the company had delivered something of a surprise in the shape of the 316i Compact, an entry-level model designed to counter what the BMW board saw as the up-market aspirations of Volkswagen with its VR6-engined Golf. The two-door hatchback Compact, 22cm shorter than the four-door sedan, had a simplified specification and previous-generation interior and rear suspension that gave it pricing in Golf territory, while its rear-wheel drive and peppy (though basic) eight-valve, 1.6-liter engine provided a fun-to-drive element that was conspicuously lacking in the third-generation Golf. BMW believed this would give it a useful defense in the battle against the big producers—except that, in the very final months of the Compact's development, BMW had itself become a volume producer with its 1994 takeover of the Rover Group. A BMW executive confessed at the press launch of the Compact that the company would probably not have remained committed

△ **After having rejected V-8 engines during the 1970s, BMW developed a new generation for the 7- and 5-Series in the 1990s. Later iterations would move to fully variable valve gear.**

to the model had it known it was about to absorb Rover; nevertheless, this oddball hatchback was steadily improved over its seven-year production span, and later 323ti versions—with their 170-hp VANOS-equipped six-cylinder engines—were entertaining to drive and did go some way to replicating the magic of the still much missed 2002 tii.

By summer 1992 von Kuenheim's reluctance to go for speculative, full-throttle growth and jump onto the boom bandwagon had proved to be a wise strategy: the German market bubble had burst, and most automakers were busy cutting surplus capacity and firing workers as sales crumbled and export markets were slow to pick up. BMW, on the other hand, had consciously kept output behind demand, and though it suffered an 11 percent fall in sales in 1993, others were hit much harder and BMW was the only German automaker to make a profit that year.

The decade was one in which profound changes were beginning to influence the whole industry, and BMW was one of the first to embrace the environmental and societal consequences of its vehicles and its activities, presenting the first

in a series of electric car concepts at the 1991 Frankfurt show and launching initiatives in recyclability and industrial energy saving. Yet it was two internal personnel developments taking place at this time, far from the public eye, that would have the biggest influence on BMW's subsequent trajectory and its world image. In October 1992 a little-known American designer joined BMW's styling studio from Fiat, where he had provoked a minor fuss with his polarizing design for the Fiat Coupe. His name was Chris Bangle, and he, together with the youthful Bernd Pischetsreider (who had joined the BMW board as manufacturing director in 1991 and was being groomed as von Kuenheim's chosen successor), would change the face and footprint of BMW forever.

117

12

The Boxer and After

For BMW motorcycles, the 1970s were a decade of challenges. After a brief renaissance following the launches of the R90S, the R100S, and the touring R100RT earlier in the decade, BMW Motorrad's momentum appeared to evaporate. The Japanese were gaining in almost every capacity class; the exchange rate was going the wrong way in the main export market, the United States; and BMW bikes were perceived as being conservative as well as expensive.

The company's planners faced a difficult dilemma: should they stay faithful to the famous boxer flat twin that had been systematically developed since the 1920s, which was so strongly identified with the brand, or should they make a clean break and invest in new and more powerful engines to match the showroom appeal of the Japanese producers' multicylinder units?

In the event, BMW did both. The classic designs are now stronger than ever and the boxer has not only managed to outlive the engine designed to replace it, but—with a power output almost fifteen times that of the 1923 original—it now proudly rides alongside singles, parallel twins, across-the-frame fours, the spectacular six-cylinder K1600, and a growing range of premium scooters.

Business has never been better for BMW Motorrad, on a roll for the past five years and selling nearly 137,000 bikes in 2015. But not every chapter of Motorrad history was

a glorious one: the brave C1 scooter-with-a-roof flopped badly at its 2000 launch and never recovered, the company's attempt at an American-style cruiser failed to hit the mark, and the romance with the Husqvarna off-road brand proved short lived. Yet these were just isolated inconsistencies in an otherwise smooth ascent, a story that has seen BMW reinforce its position as the world's premier quality motorcycle brand with premium offerings in every sector, from urban commuter to enduro and from adventure sports to supersports, tourers, and top luxury cruisers.

BMW K100, 1983, and K75, 1985

In the late 1970s work began on an all-new generation of BMW motorcycles to take over from the famous boxer that had served BMW so well but had come to be seen as old-fashioned. The K-Series that emerged from this process was very different mechanically: it retained BMW's trademark shaft drive but used automotive practice (reputedly inspired by a Peugeot engine) to shape the liquid-cooled straight-four engine, which was installed flat on its side inline with the frame. The square-shaped cam covers of the 1-liter unit soon earned the K100 (pictured next page) the nickname "flying brick," but the bike was systematically improved over the years; by 1996, having expanded to 1.2 liters for the K1200 RS, it eventually gave 130 hp—the first BMW to smash through the 100-hp advisory

threshold set by German insurance companies. Two years after the K100's debut came the three-cylinder K75, which many enthusiasts found preferable for its lighter weight and easier handling. The K100 engine would prove popular in sidecar racing, and its smoothness would allow BMW to compete on level terms with the Japanese in terms of refinement, if not yet speed or glamour.

BMW K1, 1988

To the notoriously conservative motorcycle world, the K1 (left, center) came as a huge shock. Not only was it finished in a garish combination of red and yellow, but its dramatic aerodynamic styling comprised an integrated fairing enclosing the front wheel, deep side casings running under the engine, and a blended-in tail fairing. Even to today's eyes it still looks somewhat out of place—evidence, perhaps, that many of its stylistic innovations were not taken up more widely. Its mechanical advances, on the other hand, were more lasting. The K1 had triple-disc ABS brakes and was the first motorcycle in the world to include an exhaust catalyst; the fuel-injected sixteen-valve version of its four-cylinder engine was BMW's first to hit 100 hp, and the single-sided Paralever rear-suspension swinging arm was finished in the same lurid mustard shade as the three-spoke alloy wheels. Sadly, it was a step too far for the cautious biker in the showroom; despite a switch to more conservative color schemes, the futuristic machine sold too slowly and was dropped in 1993 with fewer than seven thousand built.

BMW F-Series, 1993

The remarkable success of the boxer-engined R80 G/S and its close descendant, the R80 GS, had alerted BMW and other motorcycle manufacturers to similar opportunities for cheaper and simpler all-terrain bikes. BMW's initial reaction was to cast around for off-the-shelf solutions, and by 1993 it had secured a deal with Rotax for 650cc single-cylinder four-valve engines and Italian builder Aprilia for the rest of the bike. The F650—BMW's first-ever bike with chain drive to the rear wheel—evolved into a range of road and enduro-style bikes, culminating in the high-riding Dakar version in 2000. Manufacture had by that point been brought in house at BMW, and in 2008 a water-cooled, parallel-twin-cylinder engine of 800cc replaced the single. Confusingly, the single later reappeared as the G650 series.

120

CHAPTER TWELVE

BMW R1100 RS/RT, 1992

Seventy years after the debut of the first boxer, BMW celebrated the anniversary with a fresh generation of engines that kept faith with the original's layout but brought the motors into the modern era with single-overhead-cam four-valve cylinder heads, twin oil pumps, and chain drive to the camshafts. The heads, long a borderline zone for overheating, gained oil cooling. The engines, which now gave 90 hp, were the first to offer closed-loop catalysts and acted as a stressed element of the chassis; on the sports-touring RS and later full-touring RT versions, the chassis had the innovative Telelever front suspension to minimize dive under braking, while at the rear the single-sided

Paralever swinging arm was employed. A year after the RS, the enduro GS was fitted with the new engine, and with another hike in power in 1999 this model family continued until the next big change—to water cooling—in 2013. A relatively short-lived member of the family was the 1200C, BMW's take on a Harley-style cruiser, complete with torquey engine, adjustable backrest, saddlebags, and copious chrome.

BMW R1150 GS Adventure, 2002

Described by some as the rugged Land Rover of the two-wheeled world, the GS Adventure (top) in its many evolving guises is in fact more like the luxurious Range Rover: it has impeccable quality and a real touch of class to match its go-anywhere off-highway ability and its smooth and easy on-street manners. The first GS to gain the Adventure designation was the 1150 in 2002, fitted with the upgraded four-valve engine remapped to place the emphasis on solid, low-down power; this was the bike made famous by Ewan McGregor's and Charley Boorman's epic round-the-world trips, and the halo effect of these TV series ensured the GS has remained among BMW's top-selling motorcycle lines ever since. Upgraded to 1170cc, 125 hp, and water cooling in 2013, the six-speed R1200 GS, complete with its many electronic rider aids, continues to be the definitive machine for true long-distance exploration—no matter what kind of terrain that adventure takes in.

BMW K1200, 2004, and K1300, 2008

In the world of motorcycling, BMW has a reputation for delivering the unexpected, and this certainly was the case with the K1200, released in 2004. After much research into the optimum configuration for a high-powered sports bike, BMW surprised everyone with an inline four-cylinder set across the frame—just like almost every Japanese superbike since the Honda CB750 in the 1970s. Reassuringly, however, the rear wheel was still shaft driven. BMW's 1-liter was a four with a difference: it was compact and narrow, it sat low and tilted forward in the frame thanks to a dry sump and stacked gearbox, and, thanks to input from the BMW Formula One race team, it produced an astonishing amount of power—more than 167 hp. Chassis-specific electronics matched those in the engine for sophistication, but it was unfortunately a case of too much innovation in too short a time; early K1200s were tricky and troublesome, and production had to be delayed. But by 2009, with the assistance of outside consultant Ricardo Motorcycle, the new K1300 took over with even more power and even more sophisticated chassis electronics. The four-cylinder K had become the great bike it always promised to be, and BMW became the machine of choice in the fast sports-bike sector.

BMW HP2, 2005

Another oddity in the BMW collection, the HP2 (above, right) was a further permutation on the boxer theme—one which defied both convention and, in the opinion of many, logic. The first HP2 was a mean-looking enduro, with the

air-cooled 1170 engine tweaked up to 105 hp; its long-travel air suspension, high-set mudguards, and circuitous two-into-one exhaust gave it a distinct motocross air, but its high price was a deterrent. But then came an unexpected twist: the 2007 Sport version. Now with the revvy DOHC four-valve version of the boxer engine—the most powerful of all, at 128 hp—the Sport had a highly regarded chassis with near race-specification brakes and suspension and delivered genuinely rapid performance, even by the elevated standards of Japanese bikes. Oddball though the HP2 Sport might have been, for many it is an all-time favorite.

BMW S1000 RR, 2010, and HP4, 2012

Another unexpected move from BMW came in the spring of 2008, when the company announced that it would contest the World Superbike Championship the following year and that it would have built the required one thousand road bike examples for general sale by the end of that year. To prove the point, BMW had on hand a prototype of the bike that customers would be able to buy starting in 2010, and the daunting target the engineers had set for themselves was a figure of 190 for both the engine horsepower and the ready-to-ride weight in kilograms. Straight out of the box, the S1000 RR (below) was sensational: the 999cc sixteen-valve engine was remarkable, yet also easy to ride, while the chassis was packed with innovative electronics that included race ABS and traction control, a rain mode that limited power to just 160 hp, and even an anti-wheelie control. It broke with convention in many ways, including having chain drive, but instantly became the class of the field among road-going supersports bikes. Later chassis enhancements, including the HP4 limited edition styled on the boxer HP2, added Dynamic Damping Control—effectively the world's first active suspension on a bike—and there was a slightly milder S1000 R naked version in 2013; finally, the third iteration in 2015 upped power to a whisker under 200 and gave the rider a choice of five different engine and chassis modes. Many race-inspired bikes feel out of their element on ordinary roads, but not this one.

BMW K1600, 2011

This was the big one, the bike that would confirm BMW as the builder of the world's finest motorcycle. To be the most sophisticated and the most luxurious sports tourer on the market, the new bike (above) had to be the smoothest, too—and this meant moving to six cylinders in place of the standard four. A straight-six layout was chosen, with the engine set across the frame; however, sixes are normally too wide to allow the bike a decent angle of lean on corners, so the engine's ancillaries were placed behind the crankshaft rather than on its ends. Tuned for flexibility rather than outright power, the 1.6-liter engine gave 160 hp and a solid 175Nm of torque for genuine easy riding. Needless to say, everything else about the K1600 was designed to a similarly superior standard, and the big machine's handling and comfort immediately earned high praise. The star of the show, however, was still the engine—to use the words of one leading journal: "Zero vibration, tons of torque, a soundtrack unlike any other bike and the ability to chug along at any revs in top gear. Put simply, it's the most amazing bike engine anyone has ever built."

BMW R1200 GS and RT, 2013

Beginning with the big-selling GS enduro model, BMW's boxer lineup underwent some significant changes from 2012 onward—for that is when the flat twin-cylinder engine finally broke with ninety years of tradition to abandon air cooling in favor of the water cooling relied upon by almost every other large motorcycle. Nevertheless, the engineers retained a token measure of air cooling on the cylinder barrels simply to preserve their familiar finned appearance. The DOHC four-valve cylinder heads helped expand power to 125 hp, and the gearbox (now six speed) and transmission layout were revised to take the shaft drive along the left-hand side of the machine rather than the right. The even more rugged Adventure version of the R1200 GS appeared the following year; so, too, did the sports-touring RT (below). The new RT carried across the driveline changes and some chassis updates, adding further innovations of its own centering around its role as a luxurious long-distance cruiser—Rain and Road riding modes were standard, with the option of a more dynamic Pro mode as well as hill start control to aid maneuvering. Also optional was electronic suspension adjustment, which automatically tailored the suspension to the prevailing conditions and the riding style.

BMW C600 Sport and C650 GT, 2012

With the debacle of the C1 scooter-with-a-roof still fresh in corporate minds, BMW thought long and hard before deciding to re-enter what had become known as the urban mobility market. The chosen strategy was to go in at the top of the market, in the sector for luxury maxi scooters, where BMW felt its premium brand positioning would give it the greatest advantage; in particular, company planners insisted that the scooter should handle like a proper motorcycle, thus earning the respect of the whole two-wheeler community. The C600 Sport and C650 GT (above) that launched in 2012 were as conventional as the C1 had been radical: the 650cc twin-cylinder engine used in both operated through a CVT transmission to ensure easy "twist-and-go" riding, and, just as BMW had promised, the models' dynamics were good enough to earn genuine praise from reviewers. Just one aspect came in for criticism—the high prices. In 2014 came the C-Evolution, the first battery-powered maxi scooter from a major manufacturer, and 2016 saw a midlife facelift that increased the differentiation between the Sport and GT.

BMW R nineT, 2013

Launched in 2013 to celebrate ninety years of BMW motorcycles, the R nineT (above) was a design exercise that caught the spirit of the moment so effectively that BMW was soon persuaded to put it into production. The original intention had been to produce a classic roadster bike that represented power and purity in their most distilled forms, ensuring there was maximum scope for personalization by the owner; to this end, the electrical system was specially designed to allow elements to be added and swapped. Bodywork was kept to a minimum, with just the aluminum tank and the seat nacelle, itself removable to allow the fitting of a pillion seat; the headlamp was a classic circular unit, the normal Telelever front suspension made way for the authentic look of conventional

forks, and the air-cooled boxer engine, with 110 hp, had its exhausts curving gracefully out of the fronts of the heads, then merging under the chassis to reappear as two upward-facing megaphones on the left side of the rear wheel. It was great to ride, and everyone loved it—but BMW had more surprises still in store. In late 2015 a Scrambler version was added, with high-rise bars and a sump guard, and in 2016 came perhaps the most exciting of all: the R nineT Sport, an authentic evocation of a 1960s café racer, complete with polished alloy tank, wire-spoked wheels, and racing-style seat.

BMW G310 R, 2015

BMW had not fielded an under-500cc bike since the sweet R45 boxer in 1978, and its last small-capacity single was the 250cc R27, launched in 1960; high production costs and low-priced Japanese competition had made these models unprofitable. But by late 2016, BMW felt able to return to the smaller sector with a new single-cylinder bike—though with an important twist. The machine would be designed by BMW Motorrad in Germany but manufactured by TVS in India, thus allowing a more competitive showroom price. Unusually, the G310's 34-hp single-cylinder engine was tilted backward in the frame, with the induction system to the front; as with all BMW motorcycles, ABS braking was standard, and the low 785mm seat height and trim 158kg weight made the machine attractive to smaller riders. The initial R version (left) had blue and white bodywork intended to invoke the exotic S 1000 R sports bike.

125

13

The M Cars

Two events in 1972 would have a big bearing on the way the BMW brand evolved during the next forty years and beyond. The first and the most publicly celebrated was the unveiling of the BMW Turbo design study created by Paul Bracq, powered by a 2-liter engine from the BMW 2002 and boosted to produce 200 hp, with the potential to go even higher. The second was the formation of BMW Motorsport GmbH to further BMW activities in motorsport.

The new organization, a wholly owned subsidiary of the parent company, very quickly began to produce results, starting in spectacular style with the BMW 3.0 CSL, introduced in May 1972. The model was a so-called homologation special, sold as a production car in limited numbers to make it eligible for competing in the European Touring Car Championship (ETC). Its engine was derived from the 3.0 CS but overbored to 3,003cc, making the CSL eligible for the over-3-liter class. In 1973 its capacity was further increased to 3,153cc by increasing the stroke.

It was this final version of the 3.0 CSL, complete with front spoiler, fins running along the top of the front wings, a roof spoiler, and larger rear wing, that was finally homologated for racing in July 1973. The addition of the fins and wings earned the 3.0 CSL the nickname "Batmobile." It was a huge success, winning the 1973 European Touring Car Championship and going on to win the ETC every year between 1975 and 1979.

The glamorous mid-engined BMW M1 was another homologation special, designed for Group 4 racing but completed too late to compete. Jochen Neerpasch, then head of BMW Motorsport, got around that problem by creating a race series especially for the M1 (see below).

In 1983 BMW Motorsport won the Formula One World Championship with a Brabham BMW driven by Nelson Piquet. The road car–derived powertrain produced more than 790 hp and was developed under the stewardship of Paul Rosche. In the 1980s, BMW marketing cleverly adapted the Motorsport division as a sub-brand and the "M cars" were born: ultra-high-performance production cars wearing the exclusive "M" badge. The first of these was the sensational 1984 M635 CSi with its motorsport-derived, 3.5-liter, twenty-four-valve straight-six engine. After that the floodgates opened, and today BMW M is still producing not just thrilling sedans and coupes but SUVs as well.

M1, 1978

The very first M car, the mid-engined M1 (pictured next page), was not intended as a production road car at all: instead it was initially conceived as a racing car to compete in the Group 5 category of the World Sportscar Championship. The rules demanded the M1 be homologated in Group 4 first, however, which required the building and selling of four hundred production road-going versions. As that would take time, BMW Motorsport head Jochen Neerpasch came up with the idea of a single-make race series to get the M1 on track as soon as possible: the Procar Championship was announced for 1979, to be populated exclusively by M1s.

The M1 design was first hinted at by the 1972 BMW Turbo concept: styled by Giorgio Giugiaro's Italdesign, the

M1 looked a lot like Bracq's Turbo—especially from the front, where BMW's distinctive double-kidney grille was preserved. Lamborghini was commissioned to produce the car, but when it fell into financial difficulties in spring 1978, BMW took full control of the delayed project. Final assembly moved to the German coachbuilders Baur later that year.

The M1 was powered by the M88/1, a 3,453cc, twin-cam, twenty-four-valve, dry-sump version of the existing straight-six engine. Fed by Kugelfischer mechanical fuel injection with individual throttle bodies, the standard M88/1 produced 277 hp, but in tuned form the Procar power units developed 470 hp. In downsized (3.2-liter) turbocharged Group 5 form, the engine was expected to make more than 800 hp. The engine drove the rear wheels through a ZF five-speed manual transaxle gearbox and there was double-wishbone suspension all round.

Procar ran as a curtain raiser for selected Formula One Grand Prix races and drew in racing stars of the day such as Nelson Piquet, Niki Lauda, and Alain Prost. The championship ran for the 1979 and 1980 seasons, but as the M1 was unlikely to be successful against wider opposition and BMW had announced its intention to enter Formula One races in 1982 by supplying Brabham with engines, the end was in sight. A total

of 457 M1s had been built by the time the project ended, and the M1 quickly became a rare and desirable collector's piece.

M635 CSi, 1984

The M635 CSi, or M6 in the United States (opposite top), set the world abuzz when it was announced at the 1983 IAA in Frankfurt. Its power unit was the M88/3, a derivative of the M88/1 from the M1. The main difference was the conversion from racing dry sump to a baffled wet sump—and a slight increase in power to 286 hp at 6,500 rpm. With 329Nm torque at 4,500 rpm, the 3,453cc engine was in a league of its own, giving huge tractability and a seemingly endless stream of power delivered against a howling, spine-tingling soundtrack.

The engine and the 6-Series were a perfect match. The standard 635 CSi already accommodated a 3.5-liter straight-six engine, so no major modifications were needed to cope with the power increase. The front MacPherson strut and rear semi-trailing-arm suspension remained, but the springs were uprated by 15 percent and the ride height reduced by 10mm. Bilstein gas dampers were tuned to match, and the front anti-roll bar increased from 19mm to 25mm. Steering, by recirculating ball, was available with engine-speed sensitivity and, on later cars, vehicle-speed sensitivity. The brakes were

uprated to larger 300mm ventilated discs on the front, with four-pot calipers.

Built at BMW's Dingolfing plant, the M635 CSi differed very little visually from the standard car. Inside, there was a 280km/h speedometer and a different rev counter that carried the M logo. Even by today's standards the M635 CSi was fast, accelerating to 100km/h in 6.4 seconds and with an unrestricted top speed of 255km/h.

M535i (E28), 1985

BMW's habitual rival, Mercedes, had launched the 190E 2.3-16 "Cosworth" for the 1984 model year, breaking new ground for a company that was still known for its conservatism. Munich's response a year later was the M535i (right), a sportier version of the E28 5-Series powered by the 3,430cc M30 engine previously seen only in the 6-Series.

Given that the twenty-four-valve, 286-hp M635 CSi had appeared the year before, the M535i came as a disappointment to most enthusiasts. Launched at October's Paris Mondial de l'Automobile, the M535i was equipped with M-Technic suspension and an elaborate body kit that failed to do the E28's naturally boxy profile many favors. Although 33 hp more powerful, at 218 hp,

and with more torque than the Mercedes, the M535i also weighed 168kg more than its competitor.

That said, the cars were quite different, as was the likely customer profile. The Mercedes lacked the BMW's space and comfort and the BMW's character was more that of a high-speed cruiser than the Mercedes. In the flesh, the M535i looked lower and more purposeful than in photographs, and the body kit was more than cosmetic. The M-Technic suspension gave a lower ride height and included gas-filled shock absorbers but did not severely compromise the ride. Built at the Dingolfing plant, the M535i was a mild M car, but the M5 that would follow just a few months later was anything but.

BMW M5 (E28), 1985

While the M635 CSi looked the part with its rakish coupe lines, the first 5-Series BMW to receive the powerful 286 hp M88/3 engine—the 1985 E28 M5—looked anything but. In fact, it was the ultimate "Q" car, with less visual impact even than the much less rapid but heavily body-kitted M535i. The M5 was initially built to demonstrate what BMW was capable of, and in response to requests from individual customers, twenty-five cars were delivered to owners before the international debut at the Amsterdam International Motor Show in February 1985.

BMW Motorsport initially had the capacity to build 250 M5s per year, and at launch there was a twelve-month waiting list in Germany alone. Built at BMW Motorsport's Preussenstrasse workshops and at the newly opened facility at Garching, the M5 was a potent mix of high performance and practicality, just about matching the Porsche 944 Turbo for pace but providing the room to accommodate four people in comfort. With a top speed of 245km/h and acceleration to 100km/h in 6.3 seconds, it was certainly fast. Stiffer

springs and dampers, anti-dive technology, and a limited-slip differential gave the M5 the handling ability to match its performance, and for its day it had decent brakes with four-wheel discs and ABS.

The M5 and the M635 CSi represented a new breed of car. Until they arrived, this kind of performance and high-specification engine technology had mainly been confined to extreme two-seater supercars. Indeed, the M88 engine was originally developed for the mighty M1. These two combined that performance and image with the comfort of a sedan as well as the ease of driving and flexible power delivery that was a hallmark of the big straight-six engine. Unleashed, the same engine would reveal a different side to its character, thrilling driver and occupants with a full-blooded howl guaranteed to send shivers down the spine. If anything held this M5 back it was the E28's shape, which by then was beginning to look a little dated at a time when hard-edged profiles were beginning to give way to softness in car design.

BMW M3 (E30), 1986

BMW made it clear from the outset that neither the M635 CSi nor the M5 was intended for motorsport—and it made it just as clear that the M3 *was* a racer. The initial E30 M3 was a homologation special, with five thousand road-going examples being built and sold, in left-hand drive form only—purely to satisfy homologation requirements for Group A racing. Because of that volume, the M3 was built at the Milbertshofen plant rather than at Garching. As a racing car it would prove to be without peer, winning both the European and the German Touring Car Championships twice, as well as multiple championships throughout Europe. It is still arguably the greatest BMW Touring Car of all time.

Launched in 1986, the M3 was powered by the four-cylinder 2,302cc S14 engine, based on the existing M10 family. The big difference was the sixteen-valve cylinder head, which was a four-cylinder derivation of the M1 head. The bore was larger than the stroke, a signature feature of racing engines; maximum power was 200 hp at 6,750 rpm, but in race tune this would increase to 300 hp.

The chassis changes included increasing the castor angle by a factor of three, remounting anti-roll bars, increasing rear-spring rates, and adding a faster steering rack with a ratio of 19.6:1. The transmission was a Getrag five-speed manual with dogleg first-gear position.

BMW subsequently produced three Evolution versions of the M3; though again made only for homologation purposes, they also added to the M3's growing cult status as a road car. Just 505 examples of the Evolution I were produced in 1997. Five hundred Evolution IIs were built in 1988, with significant changes to the engine increasing the power and torque to 220 hp and 245Nm. A deeper front air dam, an additional rear lip spoiler, and front brake cooling ducts made it easy to spot.

The third Evolution, known as the Sport Evolution, was introduced in 1989 and was the most radical. The capacity was increased to 2,467cc and the engine was fitted with bigger valves and more aggressive cam profiles to give 238 hp at 7,000 rpm. The arches were wider still, there were larger front and rear spoilers, and the ride height was 10mm lower than a standard M3. Colors? Red and black only. Special editions included the 1988 Europameister, the 1989 Cecotto edition with a 215-hp engine, and, the rarest of all, the Roberto Ravaglia edition, of which only twenty-five were built.

BMW M5 (E34), 1989

With near supercar performance and an engine almost equivalent in cost to an entire BMW 316, the second-generation M5 launched in 1989 was a mouthwatering proposition.

Each M5 was hand built by just two men at BMW Motorsport's exclusive facility in Garching. The straight-six M88 engine was even more potent than its predecessors, producing 315 hp at 6,900 rpm and 360Nm torque at 4,750 rpm, partly thanks to the fact it had grown in capacity from 3,453cc to 3,535cc. For the first time, a variable-length inlet manifold helped optimize torque low down the rev range while allowing the engine to breathe at high rpm. The top speed was an electronically governed 250km/h, yet despite a curb weight of 1,533kg, the M5 could accelerate to 100km/h in 6.3 seconds.

This was the M car everyone had been waiting for—it was beautiful to look at, and the stiffer, latest-generation body technology provided a sound platform for an excellent chassis. Compared to the basic 535i, the ride height was 20mm lower, and anti-roll bars were 2mm thicker at the front and 3mm thicker at the rear. Elastokinematics in the rear suspension dialed in a little understeer during cornering to keep the car stable, and a self-leveling rear axle kept the car flat whatever the load.

The steering ratio was reduced to sharpen the response, springs were stiffened with twin-tube gas dampers to match, and the M5 rolled on handsome 8×17-inch high-tech alloy wheels. High tech? Indeed they were: each five-spoke wheel had a vented aluminum cover and inner magnesium rotor to direct 25 percent more air onto the brake discs than a conventional wheel. Body treatments were confined to front and rear air dams and discreet black side skirts.

The E34 M5 was something of a watershed for BMW. Sophisticated, rounded, and intoxicating to drive, it was probably the first car BMW could truly call "the ultimate driving machine." In 1991 the European-spec M5 was given an even more powerful engine: the 3,795cc S38 B38, producing 347 hp at 6,900 rpm.

M3 (E36), 1992

The E36 3-Series model formed the basis for the M3 the world had been expecting with the E30: an M3 with a six-cylinder engine. Launched in left-hand drive form at the Paris Motor Show in September 1992, this new M3 became available in the United Kingdom the following March but not until 1994 in the United States. Unlike the E30 M3, this was not a thinly disguised racing car—it was a road-going GT car from the outset. Ultimately, BMW would produce it as a coupe, sedan, and convertible, but the coupe was the most common. The new M3 was built at Regensburg except for a small number assembled in South Africa for that market.

Like previous M cars, the changes over standard focused mainly on the engine and chassis tuning. This first M3 six-cylinder engine was designated S50 B30, a modified version of the existing twenty-four-valve M50 straight six. This variant kept the M50's cast-iron block but increased the bore and stroke to 86mm×85.8mm and the capacity from 2,494cc to 2,990cc. A modified cylinder head was equipped with individual throttle bodies, and the compression ratio increased to 10.8:1. The engine was equipped with VANOS continuously variable valve timing on the intake side only, and a free-flow exhaust system was added. Power output was

286hp at 7,000 rpm and torque 320Nm at 3,600 rpm, enough to accelerate the M3 to 100km/h in 6 seconds and to a top speed of 250km/h.

The basic E36 was a significant step forward in the evolution of the 3-Series. The front-to-rear weight distribution was a whisker away from 50:50, even with the six-cylinder engine fitted, giving the M3 its famous poise and handling balance. The E36 was also the first 3-Series to get the multilink rear "Z-axle" pioneered on the rare and quirky BMW Z1 roadster. The ride height of the M3 was lowered by 31mm and the track increased front and rear with uprated springs and dampers, thicker anti-roll bars, and there were four-wheel ventilated discs. The E36 M3 was hailed as one of the most exciting cars of its generation and one of the best handling BMWs ever.

In 1996 European-specification cars, the 3,201cc (3.2-liter) S50 B32 engine was introduced together with a six-speed manual gearbox, lifting power to 321 hp at 7,400 rpm and torque to 350Nm at 3,250 rpm. The specification varied globally: South African cars had a lower-compression 310-hp engine; North America and Canada M3s were equipped with the 3,152cc S52 engine producing only 243 hp and 320Nm.

McLaren F1 Engine and Le Mans, 1995–1997

The fastest production car in the world is a unique accolade—and that is exactly what the BMW-powered McLaren F1 was in its day, recording an official top speed of 373km/h.

The McLaren F1 was conceived by Gordon Murray and manufactured by McLaren Cars, with exterior design by Peter Stevens. It was designed around a carbon-fiber monocoque, the first in a road-going production car, and at its heart was a very special engine designed and built by BMW Motorsport especially for the F1 at Murray's request.

Paul Rosche, in charge of engine development at BMW for four decades and credited as father of the world-beating turbocharged Formula One engines of the 1980s, came up with a new V-12 design for the engine. The S70/2 was based on the architecture of the M20 "baby" production straight-six engine that had been so successful in 3- and 5-Series sedans from 1977: it was a classic 60-degree V-12 with four overhead camshafts and variable valve timing, which was cutting-edge technology for its day. Whereas the M20 had a cast-iron block, the S70/2 was

all aluminum, with Alusil cylinder bores to keep weight to a minimum and a dry-sump lubrication system to prevent oil surge under extreme cornering and braking.

Further weight reductions came in the form of ultralightweight magnesium castings for the cam carriers and covers, sump pan, and camshaft control system housings. A bore and stroke of 86mm×87mm was a perfect recipe for achieving high revs, and therefore more power. Forged pistons and individual ignition coils completed the high specification. In road-going form the engine produced 618 hp at 7,400 rpm and 650Nm torque at 5,600 rpm. In its final form the engine was 14 percent more powerful and 16kg heavier than the original concept, but it still weighed just 266kg. The entire finished car tipped the scales at 1,138kg.

Although originally planned by Murray only as a road car, the McLaren F1's crowning glory came in the form of the racing F1 GTR in 1995 and its Le Mans win of the same year. Rules dictated that the engine breathed through a restrictor, so the race version produced less horsepower, not more, than the production engine, peaking at 604 hp. Weight was reduced by 126kg, though, making the F1 more agile, and in 1996 there was a further saving of 38kg. An F1 GTR took victory at Le Mans, with other McLarens taking third, fourth, and thirteenth.

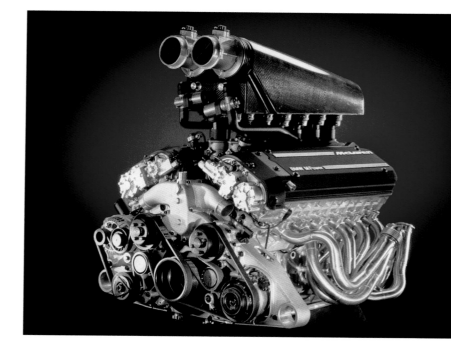

M Roadster (E36/7), 1996

By spring 1996 the M sub-brand was fully mature, and customers waited with bated breath for an M version of pretty well everything that emerged from the factory. Roadster lovers were not disappointed when, at the Geneva Motor Show in March, the Z3-based M Roadster was announced. The Z3 was assembled at BMW's Spartanburg plant in South Carolina; for the M Roadster, the drivetrain was built in Germany and shipped to Spartanburg for installation. European versions of the M Roadster were powered by the S50 B32 engine, a more advanced 3,201cc version of the M36 M3 engine equipped with double VANOS continuously variable valve timing and individual throttle bodies. Thus equipped, the M Roadster produced 321 hp at 7,400 rpm; US versions had the shorter-stroke, 3,152cc S52 engine producing 240 hp.

The M Roadster's chassis followed the E36 M3 formula but with marginally wider track, lower ride height, and stiffer springs and dampers. The subframe was reinforced and the rear semi-trailing arms beefed up too. Beginning in 1999, the M Roadster was equipped with Automatic Stability Control plus Traction (ASC+T). In 2000, European M Roadsters were fitted with a new S54 engine developing 325 hp, and this was paired with Dynamic Stability Control (DSC). In this form the car could accelerate to 100km/h in just 5.3 seconds and reach a maximum speed of 250km/h, making it one of the most potent M cars yet.

M Coupe (E36/8), 1998

An odd marriage of a roadster and what looked a bit like the back of a 3-Series Touring, the M Coupe was the personal mission of one man, engineer Burkhard Göschel, and his small group of disciples who believed the M Roadster's open body to be too flexible to do the M badge justice. Their solution was to cobble together a coupe demonstrator version and try to convince the board of directors to produce it. They succeeded, and one of the stiffest (2.6 times more than the M Roadster), most visceral BMWs ever built briefly entered production in 1998 powered by the 321-hp S50 engine for Europe and the 243-hp S52 for North America. In 2000 it was discontinued, only to be reintroduced in 2001 equipped with the 325-hp S54 engine.

The M Coupe delivered the same ferocious performance as later versions of the Roadster but with the crisper, more focused handling Göschel and his team had wanted. No one seemed sure at the time whether the E36/8 M Coupe was quirky or just plain ugly. Whatever the case, it was never a favorite with BMW marketing, and it ceased production in 2002 with only 6,291 units sold globally during its stop-start four-year life.

M5 (E39) 1998

Arguably the ultimate Q car in the history of the M cars, the 4,941cc S62 V-8-powered M5 (opposite) launched in 1998 developed 394 hp, making it the first sedan to boast 400 PS on the German industry measure. The engine was a tuned version of that powering the 540i, with the capacity increased from 4,398cc. What made it stand out from the rest—in addition to its quadruple exhausts—was the understated power delivery combined with a level of refinement that isolated the occupants from the outside world.

As a result, it was easy for the driver to become too detached from the performance, prompting one experienced journalist to comment at the international press launch that "there is the potential here to have a huge accident." He wasn't criticizing the M5's ability—far from it. The chassis was so capable and the stability so great that the M5 would accelerate to huge speeds very quickly and with little fuss, its refinement masking its velocity. That chassis incorporated the same MacPherson strut front suspension and aluminum-intensive multilink rear suspension of

the standard E39, but with thicker anti-roll bars, uprated damping, and a 23mm reduction in ride height.

The overstated engine note of the six-cylinder cars had become understated with the V-8, although the retention of a manual six-speed ZF transmission did at least help to engage the driver with the car. In numbers, that effortless performance translated to a 0–100km/h acceleration time of 4.8 seconds and an electronically limited top speed of 250km/h. The model lapped the Nürburgring in 8 minutes 20 seconds, an extremely fast time for an executive sports sedan. Large floating 345mm front ventilated disc rotors combined with 328mm for the rear to give almost unlimited stopping power, free of fade even during the most brutal track session. Unlike its predecessor, the E34 M5, this M5 was no longer hand built at Garching but made on the main production line at Dingolfing.

M3 (E46), 2000

The most potent of the M3s to date was hailed immediately after the press launch in 2000 as an all-time sporting great. In its E46 incarnation (below) the M3 was equipped with an M-Tuned S54 B32 engine, the last of the S50 line that powered the European E36 M3s. The capacity was increased to 3,246cc and the engine extensively re-engineered. An elaborate lubrication system was based around a semi-dry sump, with a second oil pump extracting lubricant from the front of the engine.

The arrangement, which included a discrete sump at the rear of the engine effectively acting as a dry-sump tank, was designed to avoid lubrication problems in high-speed left-hand bends and under heavy braking. The inlet design was substantially redesigned with help from BMW's newly re-formed F1 racing department; instead of three pairs of inlets, the E46 M3 had six individual chokes housing electronically operated butterflies but sharing a single air collector. The car now packed a powerful punch of 343 hp at 7,900 rpm and maximum torque of 365Nm at 4,900 rpm.

The suspension remained fundamentally the same design as that of the E36, but the rear axle incorporated an innovative new limited-slip differential developed by GKN Viscodrive called the M Differential Lock. The M3 could sprint to 100km/h in less than 5.2 seconds and reach an electronically limited top speed of 250km/h. Buyers could choose between a six-speed manual gearbox and the SMG II automated manual transmission. Built at Regensburg, the E46 M3 was a formidable device—but something even more potent was yet to emerge.

M3 CSL (E46), 2003

On the face of it, the M3 CSL (Coupe Sport Leichtbau), launched in 2003, was simply what its title stated—a lighter version of the M3. But in reality it was much more. Weight was reduced by 110kg by fitting a carbon-fiber roof panel, which also stiffened the bodyshell and lowered the center of gravity. Springs and dampers were specially designed for the CSL, and the steering ratio was faster than that of the standard car.

Specially built to provide a tangible link between BMW's road cars and its Formula One V-10 engine, the M5 unit (below) had forty valves and four oil pumps to keep its one thousand components fully lubricated.

The V-10-powered M5 drove through the first seven-speed SMG gearbox, which could shift gears in just 65 milliseconds and was designed from scratch as an automated manual transmission. But however impressive the new SMG might have been, US customers were not so keen: they generated enough pressure that BMW later offered a six-speed manual version for the North American market. The M5 was joined by the E61 Touring version, and both were manufactured at Dingolfing.

Despite its large capacity, the V-10 was designed around what BMW called its "high-revving concept," speeding to an impressive 8,250 rpm. In a sense, though, the E60 M5 was two cars in one: on the one hand it unleashed ballistic acceleration (0–100km/h in 4.2 seconds) with the engine screaming to the rev limit, yet it was also able to amble along at a relaxed, fuss-free canter. Such a specialized powertrain made it (along with the M6 coupe and convertible models that shared the same setup) unique, but with the steady move toward lower revs and turbocharging to improve efficiency, it is unlikely we will see anything like it again.

The car had a unique body kit with carbon front splitters to increase downforce by up to 50 percent and a distinctive trunk lid with a raised lip. Under the hood, the engine was further tuned with slightly more radical cams and a revised exhaust manifold to give an additional 17 hp, the air intake was made from carbon fiber, and the DSC system included "M Track Mode," which considerably raised the threshold at which DSC would intervene.

The CSL had no manual option and came only with the SMG II gearbox, complete with a brutal launch-control mode. The transmission was also more aggressively calibrated to shift in 80 milliseconds in its fastest mode. Once driven, the CSL could never be forgotten. The seats were firm to the point of uncomfortable, the ride firmer still, but when opened up the engine's howl summoned a rash of goose bumps. The engine produced 355 hp at 7,900 rpm and maximum torque of 370Nm at 4,900 rpm, taking just 4.9 seconds to 100km/h. Yet even more compelling than the straight-line performance was the intoxicating sensation the entire highly strung package imparted to the driver. The best M car ever? Almost certainly.

BMW M5 (E60), 2005

We saw how the E39 V-8 M5 was the ultimate Q car, combining isolationist refinement with a bewildering turn of speed, but the next M5 trumped even that, developing no less than 500 hp and 521Nm torque from its 4,999cc ten-cylinder engine.

Z4 M Roadster and Coupe (E85/86), 2006

These two cars were a continuation of the line that began with the Z3, powered by the 3,246cc S54 B32 engine introduced in the E46 M3. Externally, the Z4 family look changed dramatically from that of the Z3, the gentle curves giving way to a more dramatic, sculpted look. More visual drama was added for the 2006 Z4 M Roadster and Coupe, with deeper inset, black kidney grilles, a rear diffuser, M Double-Spoke alloy wheels, and quadruple exhaust tailpipes.

Power and torque remained exactly the same, at 343 hp at 7,900 rpm and 365Nm at 4,900 rpm; all Z4 Ms were fitted with ZF six-speed manual transmissions and the M Differential Lock. There were some changes to the chassis compared to the standard versions: the steering was quite different, the electrically assisted system of the standard Z4 being replaced by the M3's faster hydraulic rack-and-pinion setup. Like all M cars, the Z4 M was brutally fast, reaching 100km/h in 5 seconds and a top speed of 250km/h—and although it lacked the extreme wheel arches of its predecessor, the Z4 M's looks remained controversial.

M3 (E90, E92, and E93), 2007

The M3 gained a V-8 engine for the first time with the introduction of the E90 generation, first revealed in the form of the M3 Concept at the Geneva Motor Show in 2007. Three models were offered: the E90 four-door sedan, E92 coupe and E93 convertible. The sedan and coupe versions were launched later that year and the convertible the following spring.

All three were powered by the new all-aluminum S60 B40 V-8 engine, which had a displacement of 3,999cc and four overhead camshafts. Unusually, the inlet cams were driven by chains and the exhaust cams by gears. The V-8 produced 414 hp at 8,300 rpm, initially driving through a six-speed manual transmission. In spring 2008 BMW offered the M3 with a new six-speed, M DCT dual-clutch transmission capable of shifting in less than 100 milliseconds. The M DCT reduced the 0–100km/h acceleration time by 0.3 seconds and, like the previous SMG transmissions, offered both manual and automatic shifting.

Several low-volume special versions were offered. The 2009 M3 GTS Coupe developed 444 hp, weighed 136kg less than standard, and made the sprint to 100km/h in 4.3 seconds. The CRT (Carbon Racing Technology) sedan weighed 45kg less than the standard four-door but was fully loaded with satellite navigation and a high-end sound system. It appeared in 2011, the same year that the sedan was discontinued.

BMW X5 and X6 M (E70 and E71 and F85 and F86), 2009

The first M versions of BMW's "X" cars arrived late in 2009 and caused consternation among some purists who found the idea of M-badged SUVs at odds with BMW Motorsport's heritage. However, this was more than a badging exercise, and the cars were amongst the most powerful in the world in their class. Both models were built at Spartanburg using drivetrains shipped from Germany, and both were powered by a modified version of the 4,395cc all-aluminum N63 V-8 called the S63 B44O0.

The engine had revised pistons and cylinder heads, but the biggest change was to the turbocharging: the S63 was equipped with twin-scroll turbochargers, a more responsive technology than the conventional single-scroll design. With the twin-scroll, the exhaust intake of the turbocharger is split into two pathways, or "scrolls," within the turbocharger housing itself, a crossover exhaust manifold delivering synchronized pulses of exhaust to each scroll from both banks of cylinders.

The S63 also had a high-flow air-intake system. The result was an output of 555 hp at 6,000 rpm and a mighty 678Nm of torque between 1,500 and 5,650 rpm. This represented a substantial increase of 155 hp over the standard N63 engine.

The X5 and X6 were equipped with heavy-duty versions of the ZF 6HP automatic rather than a dual-clutch transmission. The transmission was tuned to deliver faster upshifts than standard, and the software prevented automatic upshifts at the redline in manual mode. The xDrive four-wheel-drive system was calibrated to give a more rearward torque bias than standard, and both the X5 M and X6 M were fitted with Dynamic Performance Control, or DPC.

DPC was a torque-vectoring rear differential that aided handling agility by pushing extra torque to the wheel with the most grip. The DSC had a special M Dynamic Mode that apportioned more torque to the rear axle and allowed the car to be driven in an oversteer attitude if desired. So while the X5 M and X6 M may not be everyone's idea of what a BMW M car should be, no one could accuse

BMW of pulling any punches when it came to performance and driveability.

In 2014 BMW launched M versions based on the new F15 X5—the F85 X5 M, and the F86 X6 M. The twin-turbo, 4.4-liter V-8 delivered 567 hp and 750Nm, and the six-speed automatic transmission was replaced by the eight-speed ZF Steptronic. Performance remained stupendous—both the X5 M and X6 M could accelerate to 100km/h in just 4.2 seconds.

1-Series M Coupe (E82), 2010

It may not be the prettiest M car, but in spirit, the oddly named 1-Series M Coupe launched in 2010 was the closest yet to the original E30 M3 in size. The similarity ended there, however, because the newer car was equipped with a version of the 2,979cc N54 straight six with twin turbochargers. This made it not only the first M car to be powered by a turbocharged engine but also the first to rely more on torque than sheer horsepower to deliver the performance.

Although giving away 79 hp to the E90 M3, it delivered more torque at a gentle canter than the best the M3 could muster.

In absolute terms that equated to 335 hp at 5,900 rpm and 450Nm between 1,500 and 4,500 rpm, with a 53Nm overboost facility taking it to 500Nm in short bursts. Gear changing could only be accomplished by a six-speed manual transmission; there was no automatic option and, flat out from rest, the M Coupe would hit 100km/h in 4.8 seconds. With a curb weight of 1,495kg, 50/50 weight distribution, and the customary limited-slip differential, that added up to a thrilling experience for the driver. The compact nature of the M Coupe made it a favorite with road testers, even though visually it was the ugly duckling of the M Coupe family.

M5 (F10), 2012

In 2012 the E60 M5, with its V-10 engine and interstellar performance, was replaced by the F10 M5, powered by the turbocharged 4,395cc S63 V-8 engine producing 553 hp and 680Nm torque. Despite this brutal power output, the new M5 failed to seduce enthusiasts in quite the same way as its predecessor. The seven-speed DCT version weighed 125kg more than the E60 SMG, and the use of synthesized engine sound through the audio system did not quite live up to the lusty bellow produced by the naturally aspirated V-10 in full song. As with the current M3 and M4, synthesized sound was needed to compensate for the lack of natural induction note, an unavoidable side effect of turbocharging.

Nevertheless, the first turbocharged M5 could accelerate to 100km/h in 4.2 seconds and reach an electronically limited top speed of 250km/h. Were this limit not in place, it could peak at 300km/h. A competition package available from 2014 increased the output to 567 hp, and a thirtieth-anniversary edition, of which three hundred were built, produced 592 hp and 700Nm torque. In that trim the 0–100km/h time reduced to 3.9 seconds. The F10 M5 was built at Dingolfing alongside the standard range.

F12 coupe and F13 convertible M6 versions were powered by the same 553-hp engine and seven-speed DCT gearbox, but thanks to a carbon-fiber roof and other measures the coupe weighed 140kg less than the M5.

M135i (F20) and M235i (F22), 2012

The M135i hatchback was launched at the Geneva Motor Show in 2012, powered by the N55 straight-six engine producing 316 hp at 5,800 rpm. It represented a new category of BMW performance cars from the new BMW M Performance Automobiles division, intended to offer a slightly less extreme version of the M-car breed in terms of feel and usability. As such, neither the M135i nor the M235i Coupe was a "full" M car. The N55 engine replaced the N54 and, although sharing the same displacement, was equipped with a single, twin-scroll turbocharger rather than twin, single-scroll turbochargers.

The M135i had a facelift in 2015 and was given an extra 6 hp to match the power of the M235i Coupe launched in 2014 with 332 hp. Both cars had either a six-speed manual or ZF's excellent conventional automatic transmission rather than dual-clutch technology. Though they were "baby" M cars, performance was strong—the M135i manual reached 100km/h in 5.1 seconds and the automatic was faster still, at 4.9 seconds. Despite their more compact dimensions, neither car was a lightweight, the M135i weighing in at 1,425kg compared to 1,520kg for the F80 M3. Built on the main 1-Series line in Leipzig, it offered enthusiasts a much lower entry-level price to the M experience.

M3 and M4 (F80 and F82), 2014

Here is where the M story becomes confusing. The M3 in its purest sense, a three-door coupe, ceased to be in July 2013 when the last E90 was produced. Its replacement, the F80 M3, only appeared as a sedan for the 2014 model year, for it was then that BMW created the 4-Series as a separate designation for the coupe, the former M3 Coupe becoming the M4.

The F80 marked a return to the straight-six engine based on the 2,979cc turbocharged N55. The S55 Motorsport version was substantially different from the standard N55. Switching to twin single-scroll turbochargers, it produced 425 hp between 5,500 rpm and 7,300 rpm and 555Nm of torque between 1,850 and 5,500 rpm. Despite being a four-door, the F80 M3 had a carbon-fiber roof and, equipped with the manual six-speed gearbox, could accelerate to 100km/h in 4.3 seconds—or 4.1 seconds when fitted with the seven-speed dual clutch transmission.

The M4, built in Regensburg, delivered the same performance as the M3 but featured extensive use of carbon fiber for the roof, prop shaft, trunk lid, and front strut brace. The M4 was also the first M car to be fitted with electric power steering, and it had Active Sound technology to amplify the engine's natural note in the cabin. The M4 may have been a far cry from the first E30 M3, but it did earn the accolade of being one of the fastest-ever cars to wear the famous M badge.

∧ **Early 2016 saw the launch of the compact M2, a car commentators felt rekindled the spirit of earlier and simpler M3s at the top of their game.**

14

Adventure
and Misadventure

The entrepreneurial 1980s were good for BMW: the deregulated financial climate in Western markets helped fuel a 50 percent jump in sales and a giant leap in its public profile; now everyone knew what a BMW was and what the brand stood for. The early 1990s were good too; the company skillfully circumnavigated the 1992–1993 recession that caused such pain to those other German companies that had expanded too rashly during the short-lived boom that followed German reunification in 1990. The new-look 3-Series was the smartest car of the moment for professionals and corporate climbers, with 1,200 streaming off the assembly lines every day; promising initiatives were also underway in almost every area, from component recycling and aero engine cooperation with Rolls-Royce to high-profile advanced engine development with McLaren for what would result in the world's fastest road car.

Reassuring though this position was—at least as far as outside observers were concerned—the forward-looking Eberhard von Kuenheim and his star young lieutenant Wolfgang Reitzle nevertheless admitted to concern about some industry undercurrents that could undermine BMW's position. Notable among these were the high costs of producing cars in Germany—which would soon be offset by the opening of a plant in the United States—and the arrival in the premium segment of Japanese firms, with their low pricing and rapid model development processes. Referring to the $40,000 price posted for the just-launched Lexus in the United States, Reitzle complained to the *Financial Times* that "this is an aggressive price: the Japanese cause us more headaches than our European rivals."

And aside from the actual cost of developing and building cars, there was another rather bigger question mark: could BMW's so-far highly successful three-pillared, sedan-centered model strategy continue to deliver the growth required to power it into the next century?

On a trip to Japan in 1987, Reitzle had suggested to von Kuenheim that BMW could enter the off-road segment—at that time just a small fraction of the market—by buying Land Rover. In a 1993 interview with *Manager Magazine* von Kuenheim explained the situation: "If necessary, we should buy their small cars [business] in the process. We wanted the off-road models,

∧ **February 1994: Rover CEO John Towers (right) seals the takeover deal with his BMW opposite number, Bernd Pischetsrieder.**

‹ **BMW's swoop on Rover took the industry by surprise and was front-page news.** *Coventry Evening Telegraph*

but we were also interested in Rover for another reason—BMW needed additional brands and models for the market below our own model series in terms of size and pricing."

With the official announcement of the forthcoming Spartanburg, South Carolina, plant in June 1992, with its seventy-thousand-unit output and 30 percent lower production costs for the affordable sports car it planned to build, von Kuenheim had begun to broaden BMW's reach. Nevertheless, it was a series of industry shocks around the turn of that year that would really jolt the company into action.

First, just before Christmas 1992, Porsche announced catastrophic losses as a result of the collapse of the American and German markets on which the elite sports car maker was highly dependent; the Porsche family members duly stepped away from the running of their company to allow a more commercial approach.

Second, incoming Mercedes-Benz CEO Helmut Werner rocked the entire auto establishment by declaring

that his company's famed products were overpriced and overengineered. "What we need are sports cars, coupés, small cars, minivans and SUVs," he proclaimed, even suggesting technical collaboration between Mercedes and BMW on component systems. Later, Mercedes-Benz would announce large losses for its passenger car operations, the first time this jewel in the crown of German industry had fallen into the red.

A third shock came in mid-March, when von Kuenheim announced that after twenty-three years in charge he would be stepping down from the chairmanship. Few had expected him to go so promptly, but fewer still had expected the comparatively unknown Bernd Pischetsrieder to be named as heir apparent: most had tipped the self-styled number-two, Reitzle, for the post, but others believed his chances had been scuppered because he had shown too much interest in a possible position at Porsche.

There could have been no better starting gun than these developments, especially Helmut Werner's *cri de coeur*, to spur the BMW board into action. The company had already decided on an acquisition, and it had the cash; the only remaining question was the target and the timing.

CHAPTER FOURTEEN

The Rover Adventure: Into the Unknown

Pischetsrieder, at just forty-five, represented a very different style of leader to his thoughtful and sophisticated predecessor. Smoking cigars, sporting a goatee beard, and professing himself fond of classic cars and fast driving, he cut a contrasting figure, prompting some editors to label him a maverick or a dark horse. But with his background in planning, quality control, and, most recently, the setting up of the Spartanburg plant, his commitment and his qualifications for the top job were not in doubt.

Presenting BMW's first-half results in July 1993, Pischetsrieder was in a confident mood despite a 9 percent fall in sales; others had been hit much harder by the recession, he said, and BMW had only had to trim production by 10 percent compared with a quarter for its rivals. Looking ahead, he wanted to improve efficiency so that the company could make money even at five hundred thousand units a year, and he even hinted that BMW could build its international position by making cars smaller than the current 3-Series, the real engine of the firm's success.

Prophetic words indeed: within a matter of months BMW officials were engaged in top-secret talks with British Aerospace, the UK defense contractor that had bought the Rover Group at a bargain price from the British government and was now looking to sell it. And it would be the decision to purchase Rover, which came as a total shock to the industry the following January, that would come to define and ultimately cut short Pischetsrieder's time in the chairman's office.

At the announcement on January 31, 1994, BMW's sense of triumph at scooping up Rover was palpable, especially as it was set against a background of dismal results from Volkswagen, Mercedes, and other German heavyweights. The £800 million handed over for priceless marques, including Land Rover, Mini, Riley, and Triumph, was widely seen as a bargain, bringing as it did a doubling of overall manufacturing capacity and four large plants with potential for much-needed expansion. Conveniently overlooked in the general euphoria was the decidedly shabby record of UK industry in quality, industrial relations, and management, as well as the inconvenient fact that Honda—which had provided the technology and the expertise that had kept Rover afloat for the past decade—had a 20 percent stake in the company and had not even been consulted about the deal.

BMW 7-Series (E38), 1994–2001

As the follow-up to the groundbreaking E32 7-Series that had drawn so many new high-status customers into BMW dealerships, the 1994 E38 came across as a conservative update on the new corporate look. Slimmer, flatter, and more discreet, it was the right car at the right time for BMW, outselling the heavyweight W140 Mercedes S-Class of the era. Mechanically, it adopted the complex multilink integral rear axle from the 8-Series coupes, and it was the first car worldwide to be offered with factory-fitted satellite navigation. It was first, too, with direct-injection diesel engines, including the 245-hp V-8 that rivaled gasoline V-8s for refinement.

1994: Launched as 730i and 740i with V-8 gasoline engines of 218 and 286 hp, with long-wheelbase versions and enlarged V-12 750i (5.4 liters and 326 hp) following later in the year.
1995: Six-cylinder 728i (193 hp) and 725tds diesel (143 hp) added.
1996: 730i V-8 replaced by 735i with larger capacity, VANOS variable valve timing, and 235 hp; 740i also gains VANOS; mild facelift to all models.
1998: Introduction of 730d with common-rail direct injection diesel giving 184 hp.
1999: New 740d is world's first V-8 diesel passenger car, offering 245 hp.
2001: Production ceases after 327,599 examples built.

BMW's outmaneuvering of its rivals to snatch the Rover prize was presented as a great coup, the cost being little more than the bill for developing a single new BMW model; Pischetsrieder would talk animatedly to reporters about his plans for a separate Mini marque, about the potential for Triumph, and about a new strategic plan for engine production spread across four countries. There were heartwarming resonances about the deal too: Pischetsrieder just happened to be the great nephew of Sir Alec Issigonis, the man who designed the Morris Minor and the groundbreaking 1959 Mini, and one of the marques in the basket of acquisitions was Austin, whose featherweight Seven baby car just happened to be the model that BMW, at that time purely a motorcycle and aero engine maker, began building under licence in 1928 to embark on its adventure as a car manufacturer.

Without question, Pischetsrieder was the motor industry's hero of the moment, and further reflected glory came when it was made known that the smart new Bentley Java concept convertible on the Rolls-Royce Bentley stand at the 1994 Geneva motor show was using BMW components. The German firm, with its eye on landing the prize of the world's most prestigious brand, had teamed up with Rolls-Royce on aero engines and was positioning itself to offer automobile technology too—a clear warning shot to Mercedes, which was also interested in supplying components as a route to subsequently acquiring Rolls-Royce. The eventual deal to supply BMW engines for next-generation Rolls-Royces and Bentleys was announced later that year—but when Vickers at last put the elite British marques up for sale in 1997, the result did not turn out quite how BMW had planned.

Though the public attention had been largely on Rover, especially the announcement that development was starting on a replacement for the classic Mini, by summer 1994 the BMW brand was up against some familiar challenges: orders outstripping production; the switchover in production to a new, third-generation 7-Series; and the spiraling value of the D-Mark, though this did bring the slight benefit of providing better value for Munich's investments in Rover.

The intention at first had been to adopt a hands-off approach to Rover, letting the British team run the operation they knew so well. Indeed, Pischetsrieder approached the Rover Group with an almost reverential respect: in an early meeting with Rover managers, he is reported to have held up two bottles of wine of different vintages, saying that it was unthinkable that these two wines should be mixed together. Interestingly, too, Rover had just been through a major reorganization of its own, a process that in parts looked to BMW as its role model; at the launch of the Honda-based Rover 600 series just a year earlier, managing director John Towers had lauded the new model as "our 4-series," the perfect fit between BMW's 3- and 5-Series of the time.

It soon became clear, however, that Rover was a much bigger mess than originally thought. Not only had chronic underinvestment by former parent British Aerospace led to a muddle of incompatible platforms and powertrains, but significant differences in work culture between the two nationalities were becoming apparent. By September, Pischetsrieder had installed the much more interventionist

Reitzle as chairman and the affable Walter Hasselkus as hands-on managing director, but the weakening of the pound sterling then began to dilute Rover earnings reaching Munich. It was the escalation of this trend that led to the United Kingdom's crashing out of the European Exchange Rate Mechanism on Black Wednesday, September 16, 1992, something that would ultimately prevent BMW from ever making a success of its Rover venture.

BMW officials continued to insist that the challenges at Rover were not having any effect on BMW itself, and this appeared to be borne out by the facts. The third-generation 7-Series that debuted in 1994 showed a discreet, elegant style

∨ The 600 was one of the few truly modern cars in Rover's range, but its Honda content posed problems for BMW.

While BMW top brass were busy dealing with Rover, the engineers continued with BMW development programs. This is the potent M Coupe, the first version of the Z3 to find true favor with sports enthusiasts.

The major commercial success of the Land Rover Freelander recreational SUV justified the decision to restrict the Discovery replacement program. ❯

that was the perfect antidote to the bulky, overstated W140 Mercedes S-Class of the day and soon began outselling it, even in Germany; the Spartanburg plant had begun production of 3-Series sedans before switching over to the brand-new Z3 sports car in mid-1995; BMW engines had triumphed in the 24 Hours of Le Mans in the McLaren F1, the first time in generations that a road-going supercar had won the grueling event; and the fourth-generation 5-Series, presented in autumn 1995, again came as a very elegant fine-tuning of the big-step 1987 design and would prove even more successful in the long-running grudge match with Mercedes. In terms of technology BMW now had the upper hand—as well as having forced Mercedes onto the back foot by being the first to offer a V-12 engine, BMW also offered superior diesels, most notably the direct-injection 3-liter in the 730d and 530d that for the first time rivaled gasoline units for refinement, and the remarkable V-8 740d that followed in 1998. The 1995 5-Series launched a further trend: saving weight and fuel by using a mixed-material structure, with the center section in high-strength steel and the front and rear—as well as the whole suspension system—in lightweight aluminum.

Uncharacteristically, there were no major new model launches from the BMW brand in 1996; instead, the emphasis was on investment to increase capacity in South Africa for the 3-Series and in Spartanburg to enlarge its scope to one hundred thousand Z3 roadsters a year. The Z3 itself was proving popular with buyers despite a guarded response from the media, who criticized its soft performance and its previous-generation rear suspension. Soon, six-cylinder engines, improved dynamics, and potent M versions would become part of the Z3 mix, and sports enthusiasts were gradually persuaded to come onboard.

Playing the International Game

BMW chiefs found themselves plunged into the world of international politics as they played off the Austrian government against the British government to gain the best possible subsidy for the large engine plant originally planned for the United Kingdom. The decision in favor of Britain was announced to great fanfare at the Paris show in October—along with another surprise announcement, that of a tie-up

BMW 5-Series (E39), 1995–2004

Like the 7-Series the year before, the fourth-generation 5-Series marked a more cautious step forward in visual design but a significant improvement in refinement, comfort, and sophistication. Employing aluminum suspension components and the clever Z-axle at the rear, the six- and eight-cylinder V-8 lineup gave standard-setting performance and handling, seizing the initiative from the Mercedes E-Class; later common-rail diesel versions set new standards for refinement and responsiveness and helped diesel become the dominant engine type in the luxury segment, while the V-8-engined M5 was the first luxury performance car to break the 400-hp barrier.

1995: Launched as six-cylinder 520i (150 hp), 523i (170 hp), 528i (193 hp), and 530i (231 hp), all with VANOS variable valve timing; V-8 gasoline models are 535i (235 hp) and 540i (286 hp), with six-cylinder 525tds (143 hp) the only diesel.
1997: Touring wagon added, featuring a two-piece tailgate and optional sliding loading floor.
1998: M5 revealed with 5-liter V-8, 400 hp, and quadruple exhausts; 530d common-rail diesel with 193 hp replaces 525tds, and gasoline models all gain double-VANOS variable valve timing.
2000: Common-rail diesel range expanded to include four-cylinder 520d (136 hp) and six-cylinder 525d (163 hp).
2001: Midlife facelift with revisions to frontal styling bringing in distinctive "corona" daytime running lights outlining the four headlights.
2003: Sedan models replaced by new E60 design.
2004: Touring ceases production, bringing overall total built to 1.49 million.

BMW's much-anticipated return to the sports-roadster market drew heavily on the previous 3-Series for its running gear and on the 1950s classic 507 for its long-hooded style. All models were built at the Spartanburg plant in the United States, but initial versions with the 1.8-liter four-cylinder engine were disappointing performers; sports enthusiasts were more convinced by later six-cylinder derivatives, with the hotshot M Coupe an oddball favorite.

1995: Two-seater roadster launched with choice of 1.8-liter (116 hp) and 1.9-liter (140 hp) four-cylinder engines.
1996: Z3 2.8i launched with six-cylinder engine giving 192 hp, followed by 3.0i versions with double VANOS variable valve timing and 231 hp from 2000.
1997: M Roadster launched with Motorsport 3.2 liter engine from M3, giving 321 hp. Distinguished by widened wheels and arches and quadruple exhausts.
1998: Z3 Coupe launched, with controversial hatchback body style; initially available only as 2.8i and, from 2000, 3.0i; M Coupe added.
2002: Replaced by Z4.

with Chrysler for the supply of low-priced four-cylinder gasoline engines to power the new-generation Mini set for a millennium debut. The investment in the engine plant—at Hams Hall in the Midlands—was part of an extravagant £3 billion package designed by Hasselkus to boost Rover's output by 50 percent and to phase in a new strategy whereby Rover would produce "relaxed rather than sporting" front-wheel-drive cars off two BMW-derived platforms (instead of Rover's current five), as well as a new Mini pitched—and priced—upmarket of current models "to reflect its cult status." Apart from the Mini, only one of these projects, the Rover 75, would ever make it into production.

Even at this stage, thirty months into the Rover project, no sharing of components, let alone engines or platforms, had been decided, and serious concerns were being voiced on both sides that the venture could run aground. A potentially even bigger concern came later in the year when the pound sterling began to appreciate against the D-Mark, a development that had the double penalties of exaggerating the latter's effects on Rover's continuing losses and making it more expensive for BMW to invest in its British arm. Ultimately, this is what would deal the final, fatal blow to the exercise.

Though Rover's losses were by now running at over £600 million a year, managers continued to insist that matters were improving and that profitability would come by 2000. In the background, Land Rover was quietly improving, and the decision to launch the car-based Freelander proved a shrewd one—even though it had drained investment cash from the much-needed replacement for the larger and mechanically outdated Discovery. Already the second-generation Range Rover was using a BMW diesel engine (in a deal made before the takeover) and a joint Land Rover–BMW team was beginning work on the much more sophisticated third-generation model. Nevertheless, BMW still saw fit to pursue its own 4x4 program, announcing that a "sports activity vehicle" would be built at the Spartanburg plant at the end of the decade. This model, the X5, proved to be an excellent insurance policy, as within a year of its launch Rover had been disposed of, Land Rover had been sold to Ford, and the outstanding BMW-inspired engineering of the third-generation Range Rover was about to be launched onto the market against BMW by a multinational competitor.

The new Mini prototype is revealed to the press in September 1997 and receives an excited welcome. Frank Stephenson (left) was credited with the exterior design.

The fourth generation of BMW's 5-Series in 1995 brought a smooth evolution of the previous style as well as lighter weight and an all-six-cylinder engine lineup. Later models added state-of-the-art diesel engines and the "corona" headlight rings that triggered today's fashion for light signatures.

154

The Range Rover development program was a rare example of harmony between the British and German engineering teams: elsewhere in the organization, the decision was made to prioritize the development of a medium-large sedan, with the Mini replacement (a program riven with dissent) given second billing and the large-volume replacements for the Honda-based 200 and 400 series still to be given a firm go-ahead.

Yet despite the escalating doubts and tensions on both sides, there was an atmosphere of real excitement at the press conference called just before the 1997 Frankfurt show. Pischetsrieder had plenty of good news to deliver: BMW's return to Formula One with the Williams team, a 30 percent group profits rise for the first half of the year, and the highest Rover sales in eight years. And then, noisily and dramatically, the new Mini prototype, resplendent in Monte Carlo Rally red with a white roof, zoomed in, its engine revving urgently. Never mind that this was in fact only a styling mockup based on a shortened Fiat Punto: the effect was electric, and all the media present could see that the replacement for the classic Mini would be racy and fashionable, prioritizing nostalgic, retro style and image above engineering innovation. The interior was especially striking, featuring a huge central speedometer caricaturing and exaggerating the design of the 1959 original. Suddenly, the sense of future potential pushed the general Rover gloom into the background and, though the Mini prototype did not appear on the show stand, the momentum was established in the media and carried on with Henrik Fisker's unashamedly retro Z07 concept—which would become the BMW Z8 roadster—at the Tokyo show the following month.

Back on the BMW front it was business as usual, with steadily rising volumes for the three main model lines and, at 1998's Geneva show, a triumphant display of the BMW brand's strength with record profits and sales, the dramatic new M5 sedan with no less than 400 hp, and the attractive, all-new 3-Series to further boost the appeal of the company's biggest selling range and its greatest earnings driver. Like the new 5- and 7-Series before it, the E46 3-Series represented a cautious update of the previous generation, tidying and refining the style rather than revolutionizing it. The move confirmed a long-suspected pattern within BMW design: that of calculatedly dramatic, era-defining jumps in style alternating

BMW X5 (E53) 1999–2006

The first 4×4 from BMW embodied "sports activity vehicle" thinking and used passenger-car suspension and steering to provide responsive on-road handling, with large tires and air-spring struts to provide ground clearance off-road. All models were built at BMW's Spartanburg facility. The initial 4.4-liter V-8 engine was steadily extended in terms of capacity and power, but in Europe the six-cylinder models proved to be the bestsellers.

1999: X5 launched with 4.4-liter thirty-two-valve V-8 (286 hp) from 5-Series; choice of five-speed automatic or manual transmission. Version with 3.0i gasoline engine (231 hp) added.
2000: Launch on European market, with 3-liter turbodiesel engine added.
2001: V-8 engine enlarged to 4.6 liters and 347 hp.
2003: 3-liter diesel uprated to 218 hp; all models receive new xDrive four-wheel-drive system and facelift to front grille.
2004: V-8 engine enlarged to 4.8 liters and 360 hp, linked to six-speed automatic.
2006: Replaced by new-generation E70 X5 after 580,000 units built.

155

Rover 75 (RD1 and R40), 1998–2005

The only new Rover design produced entirely under BMW's ownership, the 75 was the first Rover developed by largely British engineers free of budgetary constraints; many believe it was the finest car Rover ever produced. Using the BMW 5-Series Z-axle platform reconfigured for front-wheel drive, the 75 housed Rover K-Series gasoline engines as well as BMW diesels. The design, especially the elegant interior, was widely admired, and a stylish Tourer wagon derivative, developed in parallel, followed after Rover was sold.

1998: Presented at Birmingham International Motor Show.
1999: Production versions receive positive media reviews.
2000: Assembly moved from Cowley (Oxford) to Longbridge (Birmingham) following Rover Group's sale to Phoenix Consortium.
2001: Tourer station wagon and MG-branded sporting derivatives shown.
2003: MG ZT V-8, developed by MG Rover after the separation, introduced with Ford V-8 engine and rear-wheel drive.
2005: Production stops after collapse of MG Rover. Design rights sold to SAIC in China.

▲ **BMW's biggest-selling model, the 3-Series, was renewed in 1998, with the fourth generation refining and enhancing the look of its predecessor.**

with smaller and less shocking forward steps, ensuring the ideal blend of continuity—to keep residual values strong—and change, to keep the company in the vanguard of premium design. With Coupe (now designated Ci and Cd) and Touring models following within a year, the E46 was an immediate hit and straightaway set a blistering sales pace, aided in no small measure by the brilliant direct-injection 320d with the new-technology diesel engine that would become the star performer in all of BMW's model lines except the prestige 7-Series. Before long, the range expanded to include an elegant Ci Cabriolet, the razor-sharp M3 high performer, and, surprisingly, four-wheel-drive iX versions and a new Compact derivative.

BMW could congratulate itself on a perfectly pitched upgrade, with just the right amount of innovation to draw in conquest buyers as well as to keep loyalists onboard; by the time the E46 was fully replaced by the E90 models in 2006, some 2.7 million examples had streamed out of four factories to bring it level with its predecessor as the company's most successful model ever.

In the interim, Pischetsrieder had engaged in a remarkable bout of corporate brinkmanship on a German golf course with the formidable VW Group CEO Ferdinand Piëch to secure the rights to the noble Rolls-Royce marque from 2003, but the news from Rover was becoming grimmer by the day. All

> ∧ BMW takes victory in the 1999 24 Hours of Le Mans in its own car—with a little help from the Williams Grand Prix team.

> ⟩ BMW had long harbored an interest in the Rolls-Royce Motor Cars marque and was already in partnership with its aero engine division. In 1998, after Volkswagen had bought the combined Rolls-Royce Bentley group, a deal was stuck with VW CEO Ferdinand Piëch for BMW to produce Rolls-Royces from 2003.

157

The Rover 75 was the only entirely new model produced under BMW's ownership. Its graceful, classical style and classy interior were widely praised, and the engineering reflected BMW's high standards. But CEO Bernd Pischetsrieder spoiled the model's triumphant show debut by criticizing the poor productivity of Rover's UK workforce.

non-essential spending was postponed, the BMW share price was under pressure and the corporate vultures were hovering. Emboldened by Daimler's surprise takeover of Chrysler in May of that year, commentators began declaring BMW and the also-vulnerable Volvo as the next takeover targets.

Worse, what should have been a moment of triumph for Rover was turned into a public relations disaster by an outburst from no less a figure than Pischetsrieder himself. The Rover 75 sedan—the only product developed wholly under BMW ownership—had been unveiled on the Rover stand at October's Birmingham motor show to universal acclaim; its classical styling and, especially, its elegant and meticulously crafted interior wowed all who saw it, and Rover was riding high in everyone's estimation. It was unfortunate, then, that in the press conference that followed later in the day Pischetsrieder departed from his prepared script to lash out at the poor productivity of the company's UK workforce, threatening to shut down the Longbridge plant. The next morning's front pages were all about BMW attacking British workers rather than promoting the brand-new—and actually highly sophisticated—Rover 75.

Pressure on Pischetsrieder was mounting, not only from fellow board members but also from von Kuenheim, who was now in charge of the supervisory board, and members of the Quandt family. Rover was out of control, with no clear plan, and there was an impossible gulf between Pischetsrieder, who continued to profess belief in the full integration of the British volume models, and Reitzle, whose long-held view was that BMW should retain only Mini and Land Rover and sell or close down the rest. Something had to give, and, in a remarkable synchronicity of timing the following January after Ford had bought Volvo's car division, the *Economist* declared in its January 30 edition that it was "Rover's fate that will determine how Mr Pischetsrieder is judged."

The very next day, a date which would go down infamously in BMW history as "the night of the long knives," saw the longest-ever meeting of the BMW board of management, in which not just Pischetsrieder would lose his job but, less expectedly, Wolfgang Reitzle too. The man who had propelled BMW into the biggest calamity in its corporate history had failed to come up with an exit strategy and had understandably lost the power struggle that ensued. The biggest surprise was that it was not his ambitious sidekick, Reitzle, who was left to clear up the mess, but a quietly spoken and comparatively unknown former University of Munich academic by the name of Joachim Milberg.

▲ By 1999 BMW's Spartanburg plant was producing the X5 Sports Activity Vehicle alongside the Z3 sports car. The plant is now the largest in the BMW manufacturing network.

▲ Designed to drive as well on-road as off, the X5 was a breakthrough product in the SUV segment and a major success for BMW, which now has five X ranges.

◀ The retro-look Z8 roadster was directly derived from the Z07 concept car, itself inspired by the 507 of the 1950s. Lightweight space-frame construction and the M5's potent 400-hp V-8 meant it was very expensive, and it remained in production for only three years after its spring 2000 launch.

CHAPTER

15

From Disaster to Dominance

The steadily worsening condition of the English Patient—as the German press had quickly named problem-child Rover—had put BMW on the back foot throughout 1998, and at the turn of the year there was a growing sense that 1999 could prove to be the Munich company's *annus horribilis*, not just in terms of financial results but also in terms of personnel and corporate pride. By now the patient had been moved into intensive care and was being closely monitored by highly qualified specialists from Germany, but still there was division about the most sensible course of treatment.

Chairman Bernd Pischetsrieder continued to believe that BMW should invest still more in Rover to turn it into a volume brand. Engineering chief Wolfgang Reitzle, on the other hand, wanted to retain just Mini and Land Rover, and sell—or close—everything else. They were not the only people facing a dilemma: others, including many smaller shareholders, wanted rid of the whole Rover Group, though the Quandt family, as majority shareholder, kept their counsel and affirmed their faith in the board of management as well as the supervisory board, headed by Eberhard von Kuenheim.

Ultimately, however, the divisions between Pischetsrieder and Reitzle led to both being fired that fateful evening in January 1999 and to von Kuenheim appointing the neutral "quiet professor" Joachim Milberg as chairman

to restore a sense of unity. It is significant that in many of his subsequent public comments Milberg stressed that the decisions and the policies reflected the views of the *entire* board of management.

Safely installed in the chairman's office, Milberg set about taking a much firmer grip on Rover—which, arguably, should have always been under the direct control of Munich. Professing his confidence in the UK operations, he declared with disarming clarity that "the multi-brand strategy that was pursued was right in principle, but the type of group leadership proved to be wrong." The remark came as an uncharacteristic admission of failure from a company that was simply not accustomed to it. There was nothing within the BMW management culture to provide guidance once the customary consensual processes had been exhausted, and, deeper down—as evidenced by accounts from former Rover employees—there was a persistent gulf between the educated and highly disciplined German managers and their less diligent British counterparts. But now, with directors' blood fresh on the carpet at the head office, something really had to give— especially as a leaked internal report (soon denied) revealing a worst-case prediction of £900 million losses for Rover in 1999 was published in German newspaper *Die Zeit*.

The shareholders' meeting in March was understandably tense as takeover speculation swelled and wild accusations

Chairman Bernd Pischetsrieder believed Rover could prosper with further investment, but head of vehicle development Wolfgang Reitzle wanted to sell Rover but keep Land Rover and Mini. There was no compromise, so both lost their jobs and the supervisory board appointed Joachim Milberg as the new chairman. ❯

162

< The initially US-built Z4 roadster was well received and is a fine example of Chris Bangle's "flame surfacing" design technique.

∧ Joachim Milberg brought unity and stability back to BMW after Rover was sold.

were hurled at BMW's directors. A particular target was supervisory board chairman von Kuenheim, who was accused by one small shareholder of "a soap-opera performance in his mishandling of the change of CEO." Von Kuenheim retorted that it had been the right course of action to restore stability, but he nevertheless confirmed his planned resignation from the supervisory board.

Lauding von Kuenheim's thirty-year tenure at BMW and its expansion under his leadership from a regional *Mittelstand* company to a thriving global enterprise, Milberg declared that an era had come to an end. And indeed it had, in more ways than one. The man whose vision had built BMW into an

object of international admiration was no longer onboard, the architect of the company's only real wrong move had jumped ship, and the continued connection with Rover, now under a high-powered microscope, had become a dispassionate business decision rather than an article of faith. And with internal metrics showing the patient continuing to weaken, exit strategies for Rover were already under discussion: the end was in sight, and BMW could soon look forward to an end of the Rover cash drain. One or two hurdles might have been in the way, but with the entire BMW organization eager to return to its normal ethos of ambitious expansion, there was everything to play for.

In Fine Form—Apart from Rover

Away from the internal tensions surrounding Rover, BMW proved to the outside world that it was in fine creative form. The new 3-Series sedan launched the previous autumn was racing out of the showrooms at record speed, the Touring and Ci Coupe derivatives were set to push annual volumes up towards a third of a million, and the technically accomplished V-8 diesel in the 7-Series 740d was universally praised as equal to gasoline models for performance and refinement—and of course dramatically better in terms of economy and CO_2 emissions. And the BMW team won the prestigious 24 Hours of Le Mans in the BMW-powered V-12 LMR, co-designed with Grand Prix specialist Williams as an overture to the two organizations combining for a full Formula One campaign the following season.

An auspicious development was the production of the first X5 four-wheel drive at the Spartanburg plant in the summer, making good on Pischetsrieder's promise to broaden BMW's model portfolio. BMW engineers had been shocked at the clumsy handling of Land Rovers and other traditional 4×4s on paved roads and knew that a BMW for the SUV market would have to be responsive and rewarding to drive if it were to remain faithful to the brand's "sheer driving pleasure" ethos. Rather than struggling to make a trucklike

design handle acceptably on the road, the development teams had already resolved to work from a carlike platform and add off-road capability through large tires and adjustable ground clearance.

Thus the X5 was born, an amalgam of 5-Series chassis systems and 7-Series powertrains, wrapped in a tastefully smooth but nevertheless impressive station-wagon body riding high on its outsize wheels and tires. The only evidence of any Land Rover connection was in the X5's Hill Descent Control, which automatically worked the brakes to steady the vehicle's speed down even the most fearsome of off-road downgrades. Most importantly, the X5 drove with great aplomb, displaying genuine steering accuracy (until then never found on SUVs) and restricted body roll on bends. It was easy and enjoyable to drive, fully deserving the title of Sports *Activity* Vehicle (SAV) that BMW insisted on giving it in order to distinguish it from its crude trucklike competitors. Initial output of X5s was for US customers, with a diesel version added to the gasoline V-8 for European buyers the following year. Instantly successful, it launched a family of SAVs that would eventually account for one in every three BMW sales worldwide.

Industrially, the previous eighteen months had been unusually eventful. Consolidation was in full swing, with Daimler-Benz fusing with Chrysler in May 1998, then scooping

∧ From today's perspective it is hard to understand
< what all the fuss was about, but in 2001 almost every
aspect of the fourth-generation 7-Series stirred up
intense debate.

∨ A notable engineering development was the first
passenger-car V-8 diesel, which raised the bar for
diesel performance and refinement.

up Mitsubishi shortly afterwards. In January 1999 Ford had
announced it was taking over Volvo's car operations, and
Renault rescued the ailing Nissan in March to form its Franco-
Japanese alliance. Even slow-moving General Motors was
edging closer to a full takeover of its Swedish ally Saab as well
as eyeing a stake in Fiat.

To many observers, BMW, too, looked vulnerable
because of its Rover burden, so takeover talk was in the air as
1999 drew to its close. Yet for BMW, secure against predators
thanks to its stable majority shareholder, the Quandt family,
the only acquisition that could be contemplated was in the
opposite direction—for an outside organization to take over
Rover. Flinching at the likely £1.7 billion extra investment
required to modernize Rover's sprawling and outdated
Longbridge works to build the new Mini starting in 2001 and
the all-important R30 volume car planned for a few years later,
BMW managers began secretly sounding out most of the

continued on page 168

FROM DISASTER TO DOMINANCE

At the 1999 Frankfurt show there was a new name on everybody's lips. American Chris Bangle had been working in BMW's design studios since 1992, rising to become chief designer; the Z9 coupe concept unveiled at the show was the first publicly exhibited design under his leadership. The design—a big blue four-seater coupe with a smooth profile and long, upward-hinging gullwing doors—was greeted with every emotion, from bafflement to outrage, revulsion, and even outright horror. What shocked many observers was the way Bangle had taken stylistic liberties with the familiar BMW grille—now reduced to two inset sets of vertical bars— and added an inexplicable ugly rump to the otherwise smooth and attractive tail. But the Z9 also pioneered an astonishingly simple interior, a far cry from the intricate mass of switches and gauges of a regular luxury car. The transmission tunnel was home to a large metal knob: labeled the Intuitive Interaction Concept, the device claimed to operate almost all

of the subsystems of the vehicle. It, too, perplexed onlookers but would later appear on BMW production models as the iDrive—and now almost every premium car has an equivalent control.

Many commentators dismissed the Z9 as just a one-off, a flight of fancy before a return to the conservative normal. But those who took the design seriously could see the silhouette of an attractive grand tourer beneath the show-car flourishes, and after a convertible version of the Z9 appeared a year later at the Paris motor show it was clear that Bangle was intent on ushering in a new language for BMW design that was not afraid to shock or to polarize opinion, even when it came to catalogue models that customers might—or might not—want to buy.

Matters really came to a head in summer 2001 with the launch of the new 7-Series. The first production BMW to show the full extent of Bangle's thinking, the E65 has gone

CHAPTER FIFTEEN

down in history as the design that rewrote the rules of the luxury-car class, in terms of both aesthetics and function. It was a major shock at the time and prompted a further tirade of abuse, but Bangle stood his ground, and in perhaps the greatest compliment to his vision, almost every car designer from then on was influenced by his style, in particular his concept of elaborate surfacing to add reflections and intrigue to otherwise dull areas of bodywork such as the side and hood panels.

He was given a very rough ride in the press and, especially, in the then nascent arena of online comment, where the remarks were often vicious and offensive. Like many pioneers, Bangle was fighting the battles of someone whose ideas were ahead of their time. His case was not always helped by his own explanations, which came across as complex and theoretical. Despite the best efforts of his PR minders to keep his philosophizing under control, reporters often homed in on his more far-fetched notions; tellingly, one internal note uncovered in research for this book sums up a journalist's session with Bangle as "a good interview, very few 'bonkers' statements."

The iDrive was just one example of Bangle's forward thinking to gain universal acceptance, and from today's perspective it is hard to understand what all the fuss was about.

❮ The Z9 GT concept signaled a new era in BMW's design language, but it met with bafflement at its show debut.

Design director Chris Bangle (top right) was determined to bring new energy into BMW design. Adrian van Hooydonk (right) was responsible for the design of the Z9 concept and the controversial 7-Series that picked up on its major cues. ❯

continued from page 165

world's major automakers to see who might be interested. Pretty soon it became clear that there were no takers for what was seen as damaged goods—which is why BMW granted a warm reception to two entrepreneurs from British venture capital firm Alchemy Partners, who discreetly approached the German company with a plan to acquire Rover.

BMW and Alchemy maintained covert negotiations for several months, yet even with the publication in early 2000 of dire financial results for Rover—sales slumping by one-third—Milberg and the BMW board continued to profess faith in Rover's future. At the Geneva motor show in March, company executives expressly denied that Rover was for sale. "We will not walk away from this part of the group," finance director Helmut Panke told *FT Automotive World*.

These apparently sincere pronouncements served to amplify the sense of shock when, barely two weeks later, BMW announced to the world that it had agreed to sell the car-manufacturing side of Rover, as well as the Longbridge site, to Alchemy. Alchemy managing partner Jon Moulton began the due-diligence process, stating to some public hostility that his group's plan was to slim down the Rover operation and concentrate on just the MG brand. Hotly following the sale announcement was a further shock development—BMW had sold Land Rover to Ford for a healthy £1.8 billion, neatly reuniting former development head Reitzle, who had been lured to Ford, with his pet project, the sophisticated third-generation Range Rover.

Unfortunately, the die was not yet fully cast. Alchemy did not like what its due-diligence investigators found, and the group withdrew its offer; this left an open goal for the Phoenix Consortium, a populist movement centered around Rover employees, former managers, and local businessmen. By early May, following visits by union delegations and the soon-to-be-notorious "Phoenix Four" directors, a deal to satisfy both sides had been established. Yet when announced on May 9, the agreement appeared staggeringly one sided. Phoenix would pay a symbolic £10 for all of Rover's assets, the Longbridge production plant, the large stocks of unsold vehicles, and the rights to the Rover name as long as it was not sold on. A further sweetener was a restructuring loan of £500 million to make up for the fact that Phoenix had failed to find sufficient financial backing on the capital markets.

The long-awaited, born-again Mini went into production in Oxford in the summer of 2001.

Taking over from the "quiet professor," Joachim Milberg, in 2006, Helmut Panke was exuberant and outgoing. ⌄

168

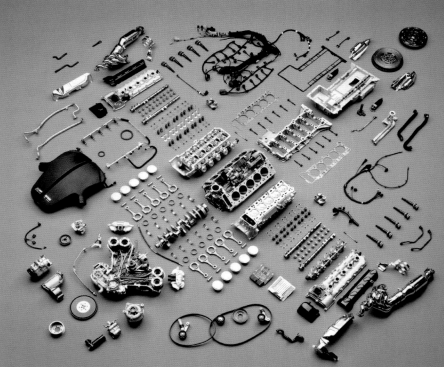

Wolfgang Reitzle with the third-generation Range Rover he initiated at BMW and launched under Ford; the 2004 M5's V-10 power unit (right) was inspired by the engines used in Formula One at the time.

Phoenix head John Towers was given a hero's welcome on his return to the Longbridge plant, and the tabloid press was jubilant that a treasured British asset had been rescued for the nation; the financial papers, however, were much more circumspect about the consortium's qualifications and the longer-term prospects for Rover as an independent automaker. BMW, for its part, was quietly satisfied that it had put the Rover episode firmly behind it and that it would not go down in history as the company that killed off the last vestiges of the once-proud UK car industry. "A luckless partnership" is how BMW's official history sums up the Rover adventure, yet at the end of the day, pride had been satisfied on all sides, BMW had freed itself from any liabilities, and Rover was free to set its own course—a course that just happened to end up in ignominious collapse just five years down the road.

BMW could at last realign its sights away from the United Kingdom and breathe again. Rover was now disappearing fast in Milberg's rearview mirror, and he was able to take in the open road ahead and resume the firm's customary foot-to-the-floor acceleration. BMW had got its mojo back, and bullish announcements flowed thick and fast: under the freshly declared "new premium" strategy, sales were running more strongly than ever before, with the half-year point showing the best six-month car and motorcycle sales in the company's history. Assembly workers' holidays were canceled, extra shifts laid on, and a thousand new employees hired to meet demand; an all-new greenfield factory was planned, with the aim of starting production in 2004.

Bucking the slumping industry trend, BMW shot back into the black, and by August its share price had jumped 50 percent since the Rover sale. The transfer of the Mini production facility to the Cowley plant—soon to be renamed BMW Oxford, to sever any lingering Morris associations—had gone smoothly, and Phoenix, now known as MG Rover, was building the well-received Rover 75 in Longbridge.

Despite heavy investment in the Mini startup, BMW was able to declare a decent surplus for 2000, and Milberg announced the start of development of "a novel model series for the upper end of the lower midrange segment within the BMW brand," as he described it in a shareholder address. This would be a VW Golf competitor to fill the gap left by the midrange Rovers, to appear on the market in 2004 as the 1-Series. Soon another development program would be added to the list—a smaller brother to the X5—to contest the market

BMW 7-Series (E65), 2001–2008

As the first production model from the Chris Bangle design team and chief designer Adrian van Hooydonk, the 2001 7-Series immediately stirred up controversy thanks to its confrontational styling and its novel methods—such as iDrive—for operating its multitude of interior systems. The car itself was large and imposing but fine to drive, with excellent engines, the world's first six-speed automatic transmission, and Dynamic Drive intelligent suspension. Successive minor exterior and interior retouches gradually took the edge off the bulky image, and the shape went on to influence design across the industry.

2001: Market launch with a choice of V-8 engines, both Valvetronic; 735i has 272 hp, 745i has 333 hp; six-speed automatic, with column selector, standard on both.

2002: Range expanded to include six-cylinder 730i (231 hp) and 730d (218 hp), 740d V-8 diesel (258 hp), as well as flagship 760i with the world's first direct-injection V-12 giving 445 hp from 6 liters. The 760i, also available with extended wheelbase, is distinguished by broader kidney grilles.

2003: Exterior facelift with smoother blending of hood and headlights (now xenon), broader grilles, and revised taillights extending inwards into trunk lid.

2005: 730i power rises to 258 hp, 735i replaced by 740i with 306 hp, and 745i becomes 750i with 367 hp. 740d engine rises to 300 hp.

2006: Limited edition of one hundred Hydrogen 7 models for selected lease customers.

2008: Replaced by new F01 model.

The fifth-generation 5-Series did not polarize opinions in the same way as the 7-Series had done in 2001. It went on to take leadership in its market segment.

segment dominated by the super-successful Land Rover Freelander, now pulling in the profits for competitor Ford. The X3 would appear in 2003, and because BMW manufacturing was pushing up against its capacity limits, it would be built by specialist contract manufacturer Magna Steyr in Austria.

The start of 2001 saw Bangle deliver another puzzling concept: the X Coupe, an asymmetric high-rider that once again ripped up the few remaining rulebooks. The macho look would resurface some years later as the X6 sports activity coupe to form yet another market niche for fashion-conscious customers. January also saw the market debut of the first of BMW's much-trumpeted personal mobility products, the C1 scooter. Equipped with a roof, windshield, wipers, and seat belts, the two-wheeler C1 claimed to be as safe as a car—riders did not need to wear helmets. But market response was lukewarm, especially as many territories required the wearing of helmets, which BMW insisted was counterproductive. Plainly ahead of its time, the C1 was a rare flop for BMW, and contract builders Bertone in Italy were instructed to cease production earlier than scheduled. By way of compensation, Bertone was given a short-run edition of a high-performance Mini—the Mini GP—in 2002.

Two key launches mark out 2001 as a pivotal year for BMW: the Bangle-styled 7-Series, coded E65, which introduced a whole

new design ethic into the large-car segment, and the long-awaited Mini. Financial analysts, habitually worried about BMW's margins, had long doubted whether the group could manufacture a Mini-sized car to its accustomed quality levels and still make a profit, especially as the initially planned volume was just eighty thousand units a year. But any doubt about the wisdom of the Mini idea quickly evaporated as Mini mania gripped the buying public, and it soon became clear that customers were ordering their Minis with an unprecedented quantity of personalizing options; in many cases, the up-spend on extra trim and equipment was as much as 30 or 40 percent of the basic list price, giving the profitability of the model a healthy boost.

BMW made a big play of the Mini's status as the world's first premium small car: it drew in buyers from all reaches of the market, being seen as the epitome of fashion and fun—especially after the arrival of the potent supercharged Cooper S version in 2002. It was not long before the Oxford plant was onto seven-day triple-shift working to meet demand. Aided by slick marketing, a regular flow of fresh derivatives such as the Convertible, and, later, the Clubman station wagon, kept Mini in the limelight, and stoked up demand worldwide so that resale values rocketed and the real cost of running the only premium small car on the market became even more attractive.

Aiming for the Absolute Summit

Ever since BMW acquired the rights to the Rolls-Royce name from VW in 1998, the date January 1, 2003, had been indelibly etched in every BMW official's diary. From that day, BMW would be entitled to use this noblest of brands and the famous double-R symbol on a production vehicle; the group intended to launch the super-luxury model—which revived the famous Phantom nameplate—at the very moment the new year dawned. The brand-new factory, set discreetly in the English countryside alongside Goodwood racetrack, had already been completed, the aluminum space-frame bodies and V-12 engines were being shipped in from Germany, and Rolls-Royce's expert craftspeople in the wood, leather, upholstery, paint, and final assembly shops were fully primed.

The first few Phantoms were completed in time for their presentation to VIPs on launch day, and it was only then that the true scale of the task accomplished by the BMW and Rolls-Royce engineers became clear. BMW,

171

BMW Z4 (E85/86/89), 2002–Present

The follow-up to the retro-themed Z3 was the first BMW to display Chris Bangle's modern flame surfacing ideas on its bodywork, creating distinctive reflections, convex and concave areas, and heightened sculptural interest. The classic roadster-style long hood and short trunk remained, but in contrast to the weedy initial Z3s, the Z4 launched with powerful six-cylinder engines, proper multilink sports suspension, and the option of six-speed SMG sequential or five-speed automatic transmission.

2003: Z4 launch lineup consists of 2.2i (170 hp), 2.5i (192 hp), and 3.0i (231 hp).

2005: Z4 Coupe concept displayed at Frankfurt motor show with fastback tail and novel flat-silver paint finish. Production is immediately approved.

2006: Z4 M launched as Roadster and Coupe, with 3.3-liter six from M3, giving 343 hp. Six-speed ZF manual transmission and M Differential Lock standard, along with a move to hydraulic power steering instead of electric.

2009: Second-generation models with comprehensively restyled body featuring retractable metal hardtop; replaces both Roadster and Coupe. Production moves from Spartanburg in the United States to the Regensburg plant in Germany. Models redesignated 23i, 30i, and 35i; latter is available with seven-speed dual-clutch transmission; others have six-speed automatic option.

2011: Engine choices revised to include 20i (four-cylinder turbo, 184 hp), 28i (four-cylinder twin turbo, 245 hp), and 35is (six-cylinder twin turbo, 340 hp).

2012: Zagato-styled coupe concept shown at Concorso d'Eleganza in Italy.

2013: Refresh of engine range adds Z4 18i, with 156 hp; all models come with EfficientDynamics revisions, and white LED "corona rings" now feature in headlights.

having acquired no more than a badge and the right to use the Rolls-Royce name, had built an aristocratic edifice from scratch, a world's finest automobile that had to perform faultlessly from the very first turn of the key; even if design director Ian Cameron's large and weighty body design initially seemed intimidating, there was no faulting the immaculate engineering and remarkable finish. Orders from the world's super-rich would soon outstrip Goodwood's limited production capacity, and, in a precise parallel with Mini, the enterprise began to expand over the decade as more models were steadily added and the factory capacity enlarged.

Away from the millionaires and back in the real world of more affordable BMWs, the new 7-Series had defied the critics and gone on to sell well since its 2001 showroom debut: customers, it seemed, were less intimidated by its monolithic shape and its next-generation electronic gadgetry than were journalists and analysts. The Z4 sports roadster, launched under incoming chairman Helmut Panke in late 2002 to replace the waning Z3, showed the Bangle design team's philosophy in a very different context. The long and low sports car broke new ground with so-called flame surfacing, the use of intricate panel shapings, and, most notably, concave sections to reflect the light and add visual interest to normally lifeless areas of bodywork. The Z4 was a particularly attractive application of this technique and still remains in production at the time of this writing—a remarkable run of over fourteen years.

The next launch was a big one—the fifth-generation 5-Series—and showed the new design philosophy best of all. Far from being the scaled-down 7-Series that many had feared, the E60/61 5-Series emerged as an attractively sculptural sports sedan with great poise and a full-length swage line that would become an even more prominent theme in later models; the Touring that followed a year later was equally accomplished in its execution. Both featured an innovative body-chassis structure blending steel and lightweight aluminum to keep the vehicle mass low and central, to the benefit of both economy and agility. The most sensational E60 was the M5, which graduated to a race-inspired high-revving V-10 of an unprecedented 507 hp, giving a top speed of 331km/h (205 mph) if the restrictor was disabled.

The fifth-generation 5-Series was BMW's first mainstream launch after the shock of the 2001 E65 7-Series, but avoided the bulky Bangle-inspired style of the larger car: instead, the E60 and its later E61 Touring derivative came across as attractively sculptural and sporty. The mechanical specification reflected BMW's engineering advances, with a blend of aluminum and steel in the structure to keep mass low and ensure agile handling. Apart from the top 545i V-8, all the engines were sixes and all transmissions, whether manual or automatic, were six-speed. Entirely new was the optional Active Front Steering. The simplified iDrive now included a color screen and the option of a head-up display for vital information.

2003: Six-model sedan range launched, spanning from 520i (170 hp) to 545i (V-8 with Valvetronic and 333 hp) gasoline and 525d and 530d diesels (177 and 218 hp).

2004: Touring (E61) models added; high-performance 535d with two-stage turbocharging and 272 hp joins range; M5 Concept unveiled at Geneva show with 5-liter V-10 engine giving 507 hp. Long-wheelbase L versions built in China.

2005: M5 launched with high-revving V-10 engine and seven-speed sequential transmission with eleven different programs.

2007: Minor facelift brings revised headlights with clear glass covers, LED rear lights with fan pattern, and revised bumpers. Powertrains gain EfficientDynamics developments, such as brake energy regeneration. Four-cylinder 520d added; other engines rise in power. M5 Touring added.

2008: Revised iDrive operation with optional hard drive and onboard Internet access; optional all-wheel drive.

2010: Replaced by sixth-generation F07 models.

^ **The most commercially significant launch of 2004 was the 1-Series. BMW believed rear-wheel drive would be a key selling point, but it was the rear-biased styling that again provoked debate. The soft silver paint finish on the 2005 concept study for a Z4 coupe (below) highlighted the sophisticated convex-concave surfacing of the body design.** ˅

BMW X3 (E83), 2003–2010

When BMW sold Land Rover in 2000, it was able to give the green light to the E83 project to produce a smaller brother to the X5, a design that would compete head-on with the successful Land Rover Freelander. But when the X3 emerged in 2003, it was almost as big as the X5, though less expensive. Like the X5, its car-based underpinnings made it good to drive, if somewhat stiff in its ride: the novel, multiplate center differential aided responsiveness on twisty roads. In its looks it lacked the classy elegance of the larger model, especially at the rear, and overall the impression—certainly until the major upgrade in 2006—was of a poor relation to the X5. Initially, all X3s were assembled by Magna Steyr in Austria.

2003: xActivity concept shown as a teaser at the 2003 Detroit show in January; production version, with 3-liter gasoline and diesel engines, launches in September.
2004: Four-cylinder engines added.
2006: Comprehensive facelift, with new front and rear lights, revised bumpers, restyled interior, new xDrive AWD system, and softened suspension. Range now topped by 35d with twin-turbo diesel six with 286 hp, and six-speed automatics now offered.
2010: Replaced by new F25 model, with production now at Spartanburg.

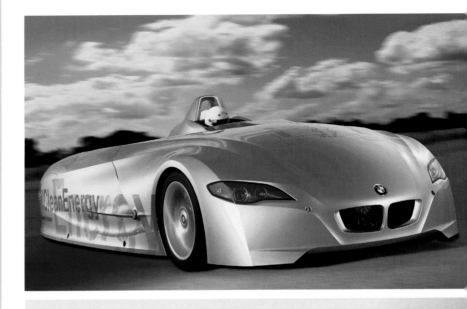

^ **BMW developed the H2R streamliner to research hydrogen propulsion systems.**

Additional releases, such as the direct-injection V-12-powered 760i, the X3 medium-sized sports activity vehicle, and the surprise return of the 6-Series luxury sports car, gave a further upward turn to the sales graph. By the end of 2004, BMW was able to announce not only another best-ever year but also a milestone its previous generation of managers could only have dreamed of—overtaking its long-standing rival Mercedes-Benz to become the world's biggest premium manufacturer. Included in the BMW Group's 1.2 million total were the first ever million-plus BMW-brand sales in a single year, plus 184,000 Minis, squeezed implausibly from a plant configured for barely 120,000, and almost 800 Rolls-Royce Phantoms. Motorcycles, separate to the car total, nudged ever closer to six-figure sales, a hurdle the company would breach just two years later.

During the year BMW's sleek H2R set speed records for a hydrogen-powered vehicle, a joint venture was signed with

 Helping celebrate seventy-five years of BMW roadsters in 2006, from left to right: Z1 (1988), Z3 (1995), Z8 (1999), and Z4 (2002).

BMW 6-Series (E63/64), 2003–2010

Reviving the spirit of the classic 635 CSi, which ended production in 1989, the BMW 6-Series is the model that most clearly shows the genes of Chris Bangle's controversial 1999 Z9 concept car. By the time of the E63's release in 2003, the furor had died down: the model was accepted as a refined and powerful grand tourer. With engineering drawn mainly from the 5-Series, the 6 was powerful and enjoyable to drive: the later addition of M versions and a world-first diesel in a luxury sports car showed the range was able to move with the times.

2003: 645Ci Coupe makes its debut at Frankfurt show, with 4.4-liter V-8 giving 333 hp.
2004: 645Ci Cabriolet revealed at Detroit show; 630i, with 3-liter six-cylinder engine giving 258 hp, announced; M6, with 507-hp V-10 engine, becomes top model.
2006: M6 Cabriolet added; 645Ci becomes 650i with larger 4.8-liter engine and 360 hp.
2007: Twin-turbo 635d becomes first luxury sports car with diesel power; 3-liter gasoline engine gains direct injection. Minor design retouches on all models.
2010: Production ceases pending introduction of new F12/F13 models in 2011.

Brilliance China Automotive Holdings for BMW manufacture, and in sweeping organizational changes, Chris Bangle took group-level responsibility for design, with Adrian van Hooydonk named chief designer for BMW and Gert Hildebrand taking direct charge of Mini design. Commercially, however, the most significant debut was that of the 1-Series, the compact five-door hatchback designed to do battle with the VW Golf and other family cars with premium aspirations.

The 1-Series stirred up many of the old arguments over so-called Bangle-style design. It was certainly controversial, its long hood and comparatively short hatchback cabin giving it a rear-biased propulsive look claimed by BMW to highlight its unique-in-segment rear-wheel-drive layout. But like most controversies, the 1-Series arguments soon died down, and the model—initially only available as a five-door—began to sell strongly to buyers who enjoyed its responsive demeanor but

The 1-Series was designed to fill the gap left after the disposal of Rover in 2000, offering an affordable model for first-time BMW buyers or graduates from Mini to BMW. Company planners believed that the "classical" BMW rear-drive layout would be a strong differentiator from the other players in the VW Golf class, but in the event it was the 1-Series' unusual styling, with a long hood and set-back cabin, that polarized opinion the most.

2002: CS1 compact Cabriolet concept unveiled at Geneva show as teaser for planned smaller BMW.

2004: Five-door hatchbacks launched, with engines ranging from 115-hp 116i to 163-hp 120d; all except 116i have six-speed gearbox, and 120i has Valvetronic.

2005: 130i introduced, with 258-hp six-cylinder gasoline power.

2007: 1-Series becomes brand ambassador for new EfficientDynamics philosophy, giving emissions reductions of over 20 percent. Three-door models added, and all models gain brake-energy regeneration and stop-start (except automatics). New 120i engine (170 hp) has High Precision Injection, and 130i with world's lightest six-cylinder block, double VANOS, and Valvetronic now gives 265 hp.

2008: Coupe model, with short sedan-style trunk intended to evoke the classic 2002, goes on sale. Engines include 120d (177 hp) and 123d (204 hp) diesels and 135i gasoline with twin-turbo six-cylinder and 306 hp. Convertible unveiled at Detroit show, 125i gasoline engine added, with 218 hp.

2009: 118d and 123d engines added; with 204 hp, 123d is world's only diesel with twin variable-geometry turbochargers, while 118d is BMW's most economical production car.

2010: 1-Series M Coupé presented, using 3-liter version of the N54 twin-turbo straight six giving 335 hp and 450Nm. Six-speed manual gearbox.

2011: Limited series of battery-powered Active E Coupes for electric-car field trials. Remaining models cease production to make way for new F20 generation.

who were not concerned about the minimal rear-seat space. BMW was insistent at the time that the 1-Series' rear-drive layout would be a strong sales draw, but in 2015, when front-wheel drive was announced for the upcoming third-generation 1-Series, customer research revealed that buyers did not care about the drive layout and many current owners did not realize their car was rear-wheel drive.

The following year, 2005, was eventful for BMW on many levels. Engineering projects begun after the Rover sale were coming to fruition, including a new incarnation of the 3-Series (which accounted for 38 percent of BMW sales); a new generation of six-cylinder engines incorporating high-precision injection and magnesium blocks for lightness; an M version of the Z4 and a stylish coupe concept; a series of small station-wagon concepts from Mini; and new four-valve engines for the evergreen boxer motorcycle. A BMW-badged minivan was mooted and a new, slightly smaller Rolls-Royce, coded RR4, was promised. On an industrial level things were gathering pace, too, with the grand opening of the Leipzig plant and cooperation agreements with Peugeot on engines and GM and Honda on hydrogen. Meanwhile, in China, the local sales company targeted annual sales of thirty-three thousand 3- and 5-Series models; by 2014, BMW's China sales were running at more than 450,000.

All this activity encouraged BMW to create twelve thousand more positions at a time when other companies were cutting back. This display of confidence once again rang alarm bells for financial analysts, who continued to complain about BMW's poor margins compared with Mercedes-Benz and a resurgent Porsche. It is perhaps significant that Helmut Panke's successor as chairman, Norbert Reithofer, began organizing road shows to brief the financial community, and that his "Strategy Number One" policy, launched in 2007, shifted the focus from unit sales to profitability and softer values such as customer satisfaction. The momentum going into 2006 was massive: advanced twin-turbo gasoline and diesel sixes, M versions of the 6-Series, a new and enlarged X5, a heavily revised X3, the second-generation Mini for autumn release, a V-12 hydrogen-powered 7-Series, diesel technology to meet strict US regulations, and the parallel-twin-cylinder F850 motorcycle. In an expansive mood delivering his final set of half-year results in August, Panke was able to step away with the cruise control set for maximum speed and a global leadership position to turn to best advantage.

Yet again BMW was faced with the task of following up the most successful act in its history—now the E46 3-Series—and the new E90 model in 2004 proved to be less controversial than the 2001 7-Series, the pronounced side swage line and the corona-ring daytime running lights standing out as its main features. Again bigger than its predecessor, the E90 had better accommodation and was also lighter, thanks to the use of aluminum components and new five-link rear suspension.

2004: Four-door sedan announced.
2005: Show debut of E90 sedan in Geneva; engine lineup includes 320d, 320i (now four-cylinder), 325i, and 330i (218 and 258 hp respectively), the latter pair sharing the same 3-liter capacity. Touring wagon version (E91) shown later in the year.
2006: Coupe (E92) added, featuring all-new body completely different to sedan, with lower roofline and broader rear lights. Cabriolet now has rigid metal folding roof instead of soft top.
2007: New M3 launched, in sedan, Coupe, and Cabriolet body styles and featuring new 4-liter V-8 engine rated at 414 hp. Initial transmissions are six-speed manual, with six-speed M-DCT added later. 335i Coupe added, with TwinPower Turbo engine giving 306 hp.
2008: Facelift gives sedan and Touring fresh bumpers, lights, hood and tailgate; xDrive AWD option added to selected models; 318d now offered with six-speed automatic and 330d now gives 245 hp, 335d is 286 hp.
2009: Improvements to all engines; new 320d EfficientDynamics edition is cleanest, most efficient BMW to date, with 109g/km CO_2 emissions. New 318d engine also has reduced emissions. Limited run of 135 lightweight M3GTS Coupes.
2010: Facelift for Coupe and Cabriolet, with reprofiled BMW kidney grilles, lights, and hood, plus interior enhancements.
2011: Announcement of new F30 model to replace E90 sedans. Runout 450-hp Carbon Racing Technologies version of M3 sedan has seven-speed DCT transmission and high-level equipment.

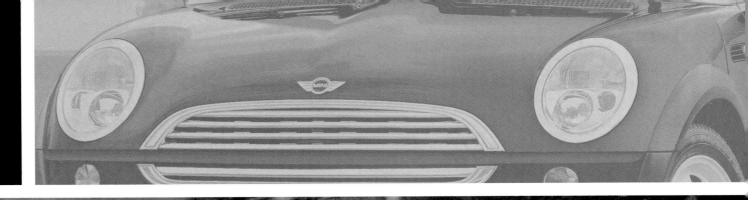

16

BMW
Unlocks
the Mini Phenomenon

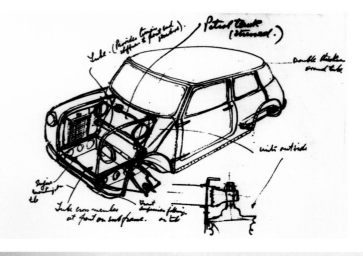

Issigonis's famous table-napkin sketch for the 1959 Mini, and the 1990s Spiritual sketch that pays tribute.

T he Mini brand has been an astonishing success for BMW and has profoundly influenced the structure of the world car market and how other global players position and promote their models. But it wasn't always so, and it nearly didn't happen at all: the Mini project had a very troubled gestation within the Rover and BMW organizations, and work threatened to grind to a halt on several occasions. The tension arose from the very different, but deeply felt, approaches of each side, the Rover contingent pushing for a technically innovative design faithful to the spirit of the revolutionary 1959 original, while BMW planners favored a more style-led theme that placed fun and fashion ahead of utility and practicality.

In line with BMW's pure-premium ethos, it was the latter approach that prevailed. Moroccan-born, US-educated Frank Stephenson came up with a funky retro design— with some input from other proposals, especially in its unashamedly outrageous interior—that was tailor made for the individualization program that would help confirm the reborn Mini's planned status as the first-ever premium offering in the small-car segment.

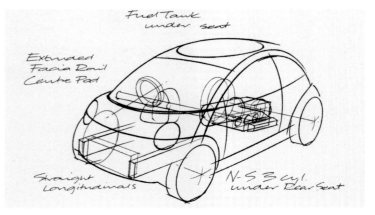

What would become the R50 program melded together a variety of elements that had been under development: a BMW front-drive platform, complete with sophisticated Z-axle rear suspension; a new four-cylinder engine co-designed and built with Chrysler in Brazil; and Stephenson's racy 2+2 body, compromised slightly to improve rear-seat and trunk space. Key to the R50's appeal was that it tapped directly into the look, the feel, and the colors of the exciting rally-winning Minis of the 1960s—even though the new design was far longer and wider and nowhere near as space efficient as the original it so closely mimicked.

Thus was born the most disruptive force the small-car scene had seen since the original Mini of 1959, a small car brimful of attitude and energy, and one whose expanding family of derivatives would preach the premium message across ever more segments to turn the whole market upside down.

Rover Spiritual Concept, 1997

Championed strongly within the UK faction of the trouble-torn Anglo-German team exploring replacements for the classic 1959 Issigonis Mini, the Spiritual revived the original's idea of truly radical thinking, with an underfloor engine at the rear and remarkable interior space in relation to its road footprint. But in the crucial 1995 shootout between the rival proposals, Spiritual lost out: BMW managers felt it did not reflect the spirit of a Mini for the twenty-first century.

BMW ACV30 Concept, 1997

The work of Adrian Hooydonk and Frank Stephenson, the ACV30 (above) was inspired by the original Mini's competition triumphs in the 1960s, reflected in its white-roofed paint scheme and its racy allure. More of a coupe than a small hatchback, the ACV30 was built on an MGF sports-car platform and was therefore rear-wheel drive. Nevertheless, it provided a clear pointer for BMW's direction of travel, as well as many of the themes for Stephenson's eventual winning design.

Mini Cooper, 2000

It was only when commentators and customers came up close to the production-specification Mini at the Paris show in 2000 that the true measure of BMW's design achievement became clear. Outside and in, this was a bold and genuinely stylish product, highly original in its scoping and its details—especially its wildly imaginative interior and dinner-plate-sized central speedometer. Most importantly, it was sophisticated and top quality in its feel, and the huge catalogue of individualization options promised each customer his or her own made-to-measure example. Never mind the high prices, the tiny trunk, or the squashed rear seating: the personalized premium small car had arrived, and the auto business would never be the same again.

Mini Cooper S, 2002

Right from the start, the pricier Mini Cooper had been the bestseller, and buyers spent a good 25 percent extra on what were, for BMW, highly profitable customization options. But while everybody adored the agile go-kart steering and roadholding, the 1.6-liter engine earned fewer compliments—until the Cooper S, that is. Fitted with a frantic, fizzing, supercharged engine endowed with a manic howl and a crackling exhaust, the S delivered bigtime performance and pulled in converts from much larger and faster cars. Rough at the edges it might have been, but the sensational S showed BMW's skill at turning up the fun factor when it wanted to.

Mini Convertible, 2004

Having a full-length convertible fabric roof yet retaining four full seats fitted in perfectly with the Mini philosophy of maximum fun for everybody. The top could be raised or lowered at the push of a button or even left in the halfway position like a sunroof; once open, the soft top rested on the rear deck to help shield rear passengers against drafts. But this Convertible was not the most practical Mini: the trunk, with its bottom-hinged lid, was truly tiny, the rear seats were cramped, and rear vision was so poor that all models came with BMW Park Distance Control as standard. Nevertheless, there was no denying its style and its customer appeal, and Convertible versions have reappeared with each successive generation of Mini.

Mini Traveller Concepts, 2005–2006

BMW had always been open about its intention to expand Mini into a family of models, and speculation about a possible wagon model had been rife. The square-backed Traveller estate car had been an important part of the original Mini's history—as had the twin, vertically split tail doors. Mini designers presented three different twenty-first-century wagon concepts at successive motor shows—Frankfurt, Tokyo, and Detroit—and each bore witness to a wild explosion of ideas for storage, seating, instrumentation, and, particularly, imaginative and hugely complex door-engineering arrangements. Taken as a whole, the Traveller series showed Mini as a spectacularly inventive brand, not afraid to take risks or abandon convention.

Mini GP, 2006

John Cooper Works (JCW) tuning packages were an important part of the new Mini's allure almost from the very start, and by 2005 the kits could be factory fitted rather than installed by the dealer. With the advent of the Cooper S, JCW kits were boosting power to 200 hp plus, but even these could not match the 218 hp of the ferocious Mini GP, a special edition that marked the end of the original supercharged S. Specially assembled by Bertone in Italy and finished in metallic gray, the two thousand examples of the GP were also lightened—mainly by taking out the back seat and most of the soundproofing— and, with their 235km/h top speed, they are prized by collectors.

Mini Hatch/Hardtop, Second Generation, 2006

By 2006 the Mini Hatch—called the Hardtop in North America—had been on sale for five years, yet its image was still vibrant and youthful and few felt it needed replacement. So BMW played it ultra cautiously with the second generation, keeping exactly the same design cues but raising the waistline, extending the front, and enlarging the lights; at the same time the speedometer grew even larger, and a few millimeters were

added in the rear seat and trunk areas. The major mechanical change was a shift to BMW's own engines, co-developed with Peugeot: the Cooper D diesel was a particular favorite, with much more urge than the weedy Toyota-sourced diesel in the first-generation model.

Mini Clubman, 2007

With a wheelbase extended by 80mm, this was the first new body style from Mini since the Convertible, and it took in some of the ideas trialed in the wagon-style Traveller show cars. The most notable—and controversial—of these was the so-called club door, a rear-hinged half-width extra door on the right-hand side, which gave easier access to the rear seat but required the front door to be opened first. In place of a regular tailgate was a pair of vertically split half doors, echoing the original "woody" Traveller of the 1960s, and the small amount of extra trunk space was welcome.

Mini Crossover Concept, 2008

The unveiling of this high-riding, four-wheel-drive concept at the 2008 Paris show marked a step change in the Mini brand. Not only was the design a lot taller and wider than the familiar Mini Hatch, but it had grown in length, too—it had now shot past the key 4m mark to give proper space for four adults. The door arrangements, always an innovative feature of Mini concepts, were exceptionally complex, with a sliding rear door on the left, a standard door on the right, and a side-hinged tailgate. The interior was novel, too, with a central rail running front to rear to carry storage systems and a speedometer that had become a glowing green laser globe. Some of the less fanciful of these ideas would even make it into the production Countryman in 2010.

Mini Coupe and Roadster Concepts, 2009, and Production, 2011–2012

While the idea of a compact Mini sports car is an appealing one, the brand's oddly proportioned concepts from 2009 did not conform to everyone's image of a stylish two-seater. The designs appeared to be little more than a standard Mini with a truncated rear, a more steeply raked windshield, and an exaggeratedly low roofline. The Coupe, with its strange helmetlike hardtop roof style, went on sale in 2011, followed a year later by the Roadster, marginally the more appealing with its curt soft top and racing-striped rear deck. Overall, however, these models were slower sellers and were pulled from the catalogue earlier than planned.

Mini Countryman, 2010

True to the promise of the 2008 Crossover concept, Mini broke out of its small-car straitjacket two years later to launch into the embryonic compact crossover market with the Countryman. On an industrial level, one surprise was that the new model would not be built at Oxford but by contract manufacturer and four-wheel-drive specialist Magna Steyr in Austria. The production car remained faithful to the concept in most aspects, though the angry frontal expression was softened and some of the wilder design touches toned down. Some ideas, such as the central storage rail in the cabin, did make it through, and though many had initial doubts about the blocky styling and overstated details, the Countryman was the first genuine four-seater Mini. It brought Mini funkiness into a mainstream sector and went on to be a major success in all its forms, right up to the top 218-hp Cooper S JCW with all-wheel drive.

Mini Paceman Concept, 2011, and Production, 2012

Another Mini to go quickly from showtime concept car to full commercial production, the Paceman entered the Mini catalogue in 2012 as the seventh body style from the brand. Many felt, just as with the Coupe and Roadster, that this was one too many and just an exercise in permutations to derive a roomy three-door fastback using the longer wheelbase platform of the Countryman. Stylistically, its only real innovation was the wedge effect of the rising waistline meeting the falling roofline to create very shallow C-pillars and a correspondingly narrow rear- window slot. Like all Mini models, it was rolled out in every guise, but this time the Mini magic was hard to find.

Mini Rocketman Concept, 2011

After the disappointment of the Paceman concept at the Detroit auto show, the Rocketman was hailed by commentators as a return to form for Mini design, by now under the direction of Anders Warming. The Rocketman re-explored the original Mini's roots as an ultra-compact car with brilliant use of space, and was just 3.4m in length—compared with 3.7m for the standard Mini Hatch. Long but ingeniously hinged side doors gave access to the interior, and the tail opening was in two parts: an upper hatch that hinged from the roof and a lower drawer that slid rearwards to aid loading and also to accommodate skis or snowboards vertically. A rich range of materials distinguished the interior, and colorful hoops on the rear flanks, much like lifting handles, served as taillights. Many hoped fervently that Mini would put the Rocketman into production, but sadly the business case was not felt to be good enough.

Mini Touring Superleggera Vision Concept, 2014

Mini's first collaboration with an outside design house led to the creation of the Touring Superleggera concept for the 2014 Concorso d'Eleganza Villa d'Este in Italy. It marked a complete break with much of what Mini had been doing up to that point: it was a dedicated two-seat sports car (in contrast to the abbreviated Mini Roadster), it had few recognizable Mini design cues apart from the mouthlike grille and the circular forms of the lights, and it was powered by electricity. Rounded and simple in the manner of the old Austin-Healey Frogeye Sprite, its bodywork was elegant, too—especially the chopped-off tail with the central fin on the rear deck and the Union Jack taillights set into the rear panel.

Mini Hatch/Hardtop, Third Generation, 2014

The look may have remained the same, or very nearly the same, but the third generation of the core Mini Hatch marked a shift to very different engineering underpinnings—the UKL platform, shared with a new generation of front-drive small-to-medium BMW models. New, too, were efficient modular three-cylinder engines for all but the most potent Cooper S and Cooper SD, which used four-cylinder motors made from the same modular toolkit. The most obvious style changes include the longer, more horizontal hood, the even bigger headlights, and the much squarer rear lights, but the big news was the appearance in 2015 of a five-door version, the result of a 161mm stretch in length that played uncomfortable tricks on the familiar Mini Hatch proportions.

Mini Clubman Concept, 2014, and Production, 2015

After the preceding incarnations of Minis with their great visual intensity and exaggerated detailing, the smooth and calm lines of the new Clubman came as a welcome contrast. It was clearly recognizable as a Mini, but, as a significantly larger car targeted at the VW Golf segment, it needed to project a more mature and less confrontational impression. The familiar Mini hallmarks were presented in a new way: the split rear doors set with horizontal rear lights and a very stylish solid-metal door pull, the big central instrument as part of a plush, luxurious dashboard. Mini had grown up. The Clubman had become a responsible family car with a dash of fun, but it was not dominated by fun to the exclusion of practicality. And, best of all, the new chassis delivered something entirely new for Mini—a halfway comfortable ride.

191

CHAPTER

17

The Finest Car in the World

Rolls-Royce and its companion marque, Bentley, had an increasingly difficult time throughout the 1970s and 1980s as their exalted standing in popular culture as the finest cars in the world began to wear thin. Their parent companies, initially aero-engine maker Rolls-Royce plc (a separate company to the automotive firm) and then defense group Vickers, struggled to find investment cash to enable the still-noble marques to match the technical development budgets of their only real competitor, the resource-rich Mercedes-Benz. By the 1990s Rolls-Royce managers knew the only way they could afford to meet ever-tightening emissions standards was to buy off-the-shelf powertrains from another manufacturer.

BMW's existing links with Rolls-Royce plc gave it a psychological advantage in securing a deal to provide V-8 and V-12 engines for the upcoming new generation of Rolls-Royce Silver Seraph and Bentley Arnage models, due in 1998. BMW believed it was in pole position to acquire Rolls-Royce Motors when it was officially put up for sale in 1997 but was shocked when the bidding war between several groups took off and was unexpectedly won by Volkswagen and its ambitious CEO, Ferdinand Piëch. Soon after the surprise deal was landed, however, it emerged that Rolls-Royce plc would need to give permission for the double-R brand name to be used in automobile applications, and that Volkswagen would not necessarily be granted that permission.

What followed was the now famous high-stakes meeting on a German golf course between Piëch and then BMW chairman Bernd Pischetsrieder, during which Rolls-Royce and Bentley—which had been joined at the hip for some fifty years—were secretly carved up. VW would take the Bentley brand, the entire worldwide dealer network, and the factory complex at Crewe where both marques were being built; BMW would be allowed to use the Rolls-Royce brand from 2003 onwards. Effectively, all BMW had bought was a name and the goodwill associated with it: no personnel, no models, and no facilities figured in the deal.

BMW was thus faced with the task, a daunting one even for its world-class design teams, of engineering from a clean sheet of paper what hard-to-please, super-rich buyers around the globe would naturally expect a Rolls-Royce to be: the best car in the world. An undercover design office was set up in London's Mayfair district to research the habits and the preferences of the world's wealthiest buyers, and after much internal debate, it was agreed that for many potential customers the archetypal Rolls-Royce was the Silver Cloud of the 1950s and 1960s. BMW planners homed in on the tall and graceful stature of the Cloud, its temple-like radiator, its sweeping lines, and its tapering tail as being the most important touch points for a new Rolls. In off-the-record briefings to selected journalists, BMW executives made it clear that sheer size and, in particular, an elevated driving position would be key qualities of the reborn Rolls-Royce generation.

As the January 2003 release date inched closer, it became clear that the new car would be of advanced all-aluminum construction, that it would have a new and sophisticated engine unique to the Rolls-Royce brand, and that it would indeed be a contender for the title of best car in the world. But of the aesthetic style being developed by Ian Cameron there was precious little to go on—until that day in January when the shape of BMW's ambition for the Rolls-Royce became clear.

Phantom, 2003

Early indications had hinted at something large and splendid, but when RR1 was finally unveiled and its Phantom name announced on January 1, the scale and the forcefulness of the design exceeded even the most extreme of expectations. Everything about the nearly 6m limousine was grand and imposing, especially the bold, bluff front with its massive grille and multiple lights; the sheer stature of the design was initially

194

intimidating, but after a while the intrinsic elegance of the lines, most particularly around the tail, could be more readily appreciated. Most vitally, though, the Phantom had achieved its main aim: sheer impact, and being utterly unmistakable in a crowd, whether in Beverly Hills or Birmingham. Backing up that visual splendor was the finest BMW-inspired engineering, with the promised new V-12 engine (of 453 hp), as well as impeccable craftsmanship in the interior and ingenious design of details marrying new technology with aristocratic discretion: rear-hinged "coach" doors to give dignified access to the back seat, the satellite navigation screen that disappeared behind an exquisitely veneered wooden panel, the weighted Rolls-Royce logos on the wheel centers that returned to vertical when the Phantom stopped, and the "Power Reserve" meter taking the place of a rev counter in the instrument panel.

100EX Prototype, 2004

Other manufacturers would have labeled it a concept car, but at the 2004 Geneva motor show Rolls-Royce took much

pride in describing the 100EX as an experimental car, in the company's finest tradition. And the 100EX was indeed a very fine example—a stylish, convertible derivative of the Phantom and full of fascinating detail that set the design world abuzz. The immediate impression of the 100EX was of something long and graceful, toning down the severity of the Phantom sedan. Among its many striking innovations were the long, polished aluminum hood, fanning rearwards from the grille to the windshield, the rear-hinged doors giving easy access to both front and rear seats, and the beautiful boat-deck-style teak planking that was exposed when the roof was lowered. The nautical theme was repeated in several areas of the vehicle, even in the trunk, which opened to reveal a teak floor and a foldout lower panel for sitting on. Further delights resided under the huge bare-metal hood: a massive 9-liter engine with no fewer than sixteen cylinders. This power unit would never feature in a production car, but almost everything else on the 100EX was applied to the Phantom Drophead Convertible launched three years later.

195

Phantom Drophead Coupe, 2007

Some had speculated that Rolls-Royce would revive the Corniche nameplate for the production version of the 100EX convertible, but at the 2007 Detroit show the model was announced as the Phantom Drophead Coupe. "This is a less formal representation of classic Rolls-Royce design," commented chairman and chief executive Ian Robertson—an observation borne out not only by the smoothly flowing lines, whether the soft top was up or down, but also by the sumptuous, full four-seater interior, with its immaculately crafted veneers, hides, and polished metal and plush flooring. The 100EX's key features, such as the rear-hinged doors and the elegant teak decking on the soft-top cover, were all retained, and at the rear the lower trunk lid folded downwards to provide a seating platform and the trunk floor lifted up to reveal a champagne cooler, six glasses, and, as an option, a magnificent cocktail basket (right). It was hard to imagine a more splendid form of travel.

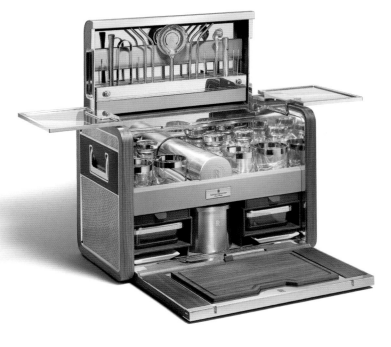

Phantom Coupe, 2008

Though outwardly similar to both the 2006 101EX concept car and the just-launched Drophead model, the Phantom Coupe marked a very subtle change for Rolls-Royce. While sharing the other Phantoms' V-12 engine and six-speed automatic transmission, the Coupe took an important step in being more deliberately focused on the driver. The transmission was equipped with a "sport" program—the first ever on a Rolls-Royce—and firmer springs and dampers, along with stiffer anti-roll bars, served to tighten up the handing and to provide a more responsive feel for the driver. Rolls-Royce chief designer Ian Cameron said, "With this car we wanted to emphasize the performance potential of the 6.75-liter V-12 engine and the effortless experience that it delivers. Whilst never overtly sporty, Rolls-Royce is a brand that has always offered owners a powerful and refined drive combined with, as Sir Henry Royce described it, a bit of fizz." As with the Drophead, the polished metal hood was an option taken up by almost every early-adopter customer, and a Starlight Headliner, threaded with thousands of tiny light fibers, was offered—just to ensure Coupe passengers did not miss out on experiencing the night-time sky.

Ghost, 2009

A smaller Rolls-Royce had always been in the plan, and in summer 2009 project RR4 was announced as the Ghost. The design interpreted Rolls-Royce visual values in a fresh and more dynamic way, with a lower and more planted stance on the road and a less imposing overall impression. The famous grille was slightly swept back, as on the Coupe, and the rear was distinguished rather than grand, but the now-familiar Rolls-Royce trademarks were carried across—the "coach" doors at the rear, the broad C-pillar to give privacy to rear-seat passengers, and of course the unmatched workmanship of every interior detail. Under the hood, however, there were big changes: the new 6.6-liter, twin-turbo V-12 was the first ever for Rolls-Royce and, with 570 hp, became its most powerful too. An eight-speed automatic transferred power to the rear wheels, and the state-of-the-art air-suspended chassis incorporated active roll stabilization and a lift-and-kneel function to aid passenger access and deal with rough roads. The suspension was said to be so sensitive that it could detect the movement of a passenger from one side of the rear seat to the other and compensate accordingly.

Wraith, 2013

Described by Rolls-Royce as the most powerful and dynamic model in the company's history, the Wraith marked a further step towards a more driver-focused car and was, with its sweeping fastback rear, the first to adopt a clearly sporting silhouette. Considerably shorter and lower than the Phantom Coupe, its hunched poise, its implicit athleticism, and its interior environment were more sporting too: the decor reflected the ambience of an exclusive yacht rather than a smoky gentleman's club, with pale, straight-grained wood paneling to the doors and dashboard rather than plush leather and dark burr walnut. On the center console the discreet cut-glass crystal Rotary Controller appeared—Rolls-Royce's interpretation of the BMW iDrive—to control car system settings. Yet surprisingly for a sports-flavored model, there was no manual override for the eight-speed automatic, the instruments still did not include a rev counter, and the gear selector still remained on the steering column, as on all Rolls-Royce models. But importantly, the 632 hp and 800Nm torque from the V-12 engine gave performance never before experienced in a Rolls-Royce, and satellite linking of the transmission allowed the driveline to look ahead and adjust its settings in accordance with upcoming conditions.

Dawn, 2015

Conspicuously lower, sleeker, and more clearly seductive than the Phantom Drophead, the Dawn (below and opposite top) was Rolls-Royce's response to the challenge of building a super-luxury convertible that was capable of accommodating four adults with no compromise whatsoever—in stark contrast to the few other competitor vehicles in the segment, most of which could only offer 2+2 seating, or worse. Giles Taylor's softer design language placed a new emphasis on gentle,

flowing curves, most clearly visible around the nose and on the rear haunches, now drawn up slightly to suggest propulsive energy; when raised, the smoothly blended-in soft top kept the side profile sporty and low. The roof itself was a work of technology and craftsmanship, special French seams allowing an unbroken surface and an almost total elimination of wind noise to achieve similar refinement to a fixed head coupe; every conceivable luxury—and more—promised to make life aboard the Dawn a hedonistic delight. Though similar in dimensions to the Wraith fastback, the Dawn was specified with a slightly, less powerful version of the twin-turbo V-12, in line with its more relaxed mission.

Project Cullinan, 2018, and the Future

In February 2015 the chairman and CEO of Rolls-Royce took the highly unusual step of issuing an open letter "on the subject of a new Rolls-Royce." In the letter, Torsten Müller-Ötvös confirmed that the company was developing an all-new model with "exceptional presence, elegance and purpose." The letter went on to explain that the car would offer the luxury of a Rolls-Royce in a vehicle that can cross any terrain, and that it would be a high-bodied car with an innovative, all-aluminum architecture. In other words, it would be a super-luxury

SUV, though no mention was made of either sport or utility: instead, the promise was of effortless luxury. Soon afterwards, the company gave a code name to the program—Project Cullinan—and revealed that a new four-wheel-drive suspension system was being developed that would deliver ride comfort that was "effortless everywhere," including off-road. The following January, the company announced that it had begun road testing the new aluminum space-frame architecture set to underpin all future Rolls-Royce model lines arriving in the market from early 2018. Müller-Ötvös closed his open letter with a well-known quote from company founder Sir Henry Royce: "When it doesn't exist, design it."

18

Powering
through the Downturn

By anybody's standards, BMW was on a roll as 2005 morphed into 2006. With the troubled Rover years now behind it, the company was entering its third year of global leadership in the premium sector; its most prestigious adversary, Mercedes-Benz, was struggling valiantly to rescue its image after electrical gremlins had given the E-Class a disastrous debut; and BMW's 1-Series, X3 SUV, and new 3-Series had helped fuel a 10 percent rise in sales to make 2005 the firm's best-ever year. The currency shifts and material costs that had rocked other automakers barely dented BMW's performance, a cost-sharing engine production deal had been signed with PSA Peugeot Citroën, and key new models on the stocks—such as the Z4 and X6—as well as an expansion in production capacity for the super-successful Mini, made for a highly confident outlook.

By the time of the London Motor Show that same summer, CEO Helmut Panke predicted a long-term growth rate of 5 to 7 percent in BMW volumes, and with his 2008 target of 1.4 million sales already within sight for 2007, he upped the stakes to 1.6 million by 2010—a goal that would only be eventually postponed because of the financial crisis of 2008–2009. So to outsiders it seemed a cruel fate that the extrovert and animated Panke had to step down at the mandatory age of sixty and that he should not be able remain onboard to enjoy the considerable product, manufacturing, and financial momentum he had helped generate during his four years at the helm.

∧ **BMW was on a roll when Norbert Reithofer took over as CEO in 2006. He planned further expansion but also prepared crisis scenarios.**

‹ **Style statement: the 2007 CS Concept previewed a new design language for BMW.**

201

▲ **EfficientDynamics was an astute move by BMW: it cut consumption but kept the fun-to-drive factor.**

Yet with today's hindsight and the knowledge of the global financial crisis that was about to come, the BMW Supervisory Board's choice of manufacturing director Norbert Reithofer as the next board of management chairman can be seen as a smart move. With his in-depth knowledge of manufacturing and production systems, Reithofer had already introduced far-reaching reforms to the group's manufacturing network, pioneering a flexible production system that used working-time accounts for each employee to ensure plants remained close to full capacity and short-term layoffs and redundancies could be avoided; it was largely this flexibility, allied with equally agile decision making, that would help BMW emerge from the crisis in much better shape than its competitors.

In the face of adverse movements in commodity prices and currencies, such as the dollar and yen, Reithofer signaled a closer focus on China, where the China-specific extended-wheelbase 5-Series was about to launch. Nevertheless, Reithofer's tone in this third-quarter presentation in late 2006, his first to the business community as chairman, was more measured and less expansive than those of his immediate predecessor: with typical caution he warned that 2007 would show a slower start because of model introductions later in the year, most notably the three-door version of the 1-Series and the high-status fourth-generation M3 Coupe. Characteristically, too, Reithofer underplayed the true potential of a perfectly timed BMW innovation that was to have a profound effect on the whole industry. EfficientDynamics was presented as a suite of relatively minor and straightforward technologies that, in combination, allowed BMW cars to be much more fuel efficient yet retain the sporty drive for which they were celebrated.

"As we see it," said Reithofer, "the main point is that a future BMW vehicle should still offer better performance characteristics than other vehicles. Sheer driving pleasure and a reasonable use of our natural resources are not opposing choices. And BMW EfficientDynamics leads the way towards this goal. You see: BMW thinks about the future!"

One of the simplest but most ingenious features of EfficientDynamics was an electronic control that disconnected the engine's alternator and air-conditioning compressor when the car was accelerating, thus avoiding power-sapping drag on the engine, but reconnected them when slowing down to help with the braking effort. Engineers and managers from other automakers were unanimous both in their respect for BMW's initiative and their fury that they had not thought of the idea first.

The timing of EfficientDynamics, whether by accident or design, proved to be a stroke of brilliance, coming as it did against a background of steeply rising fuel prices and growing concern about the impact of CO_2-based vehicle taxation then under discussion. Competitors hurried to develop their own fuel-saving solutions, leading to a profusion of green-sounding badges for almost every shape and size of vehicle. Nevertheless, BMW remained those crucial few steps ahead of the game, and by the end of the year, the company had extended EfficientDynamics from the initial 1-Series to all of its model series, the benefits being equally applicable to large luxury lines as to headline-hunting CO_2-sparing compact models.

While EfficientDynamics showed BMW with a primarily European engineering focus, the company's then-startling choice of the April 2007 Shanghai auto show for the unveiling of its extravagant CS concept revealed it as a global player with a keen eye for the likely trajectory of fast-developing new markets, as exemplified by China. Adding to the shock value was the fact that the CS, trailed as a first glimpse of the brand's new luxury-car identity, proved not to be the large and stately limousine that most had been expecting, but a sleek and sportingly low-set four-door coupe, predating the Audi Sportback and Porsche Panamera that would appear in the years that followed.

BMW executives dropped hints that the CS could be seen as a blueprint for a future Gran Turismo series, a luxury line that could become BMW's flagship product on world markets. In that role it was a clear indicator of the general sense of optimism in the premium markets—yet just eighteen months later, in autumn 2008, it was to serve as a bellwether for a very different set of economic conditions when it was canceled as the fallout from the Lehman Brothers' bank collapse spread around the globe.

Though there was little sense of concern as he presented BMW's second-quarter results early in the summer

BMW X5 (E70), 2006–2013, and BMW X5 (F15), 2013–Present

The second-generation X5 (E70) was launched in 2006 as a skillful follow-up to the 1999 original, whose sporty, carlike driving characteristics had made it a much bigger hit than even BMW had anticipated. It grew considerably in size and stature, to gain some distance from the midmarket X3, as well as space in the rear for a third row of seats. Engines were the familiar straight sixes—both diesel and gasoline—and increasingly powerful V-8s. The third-generation X5 (F15) took over in 2013, with reduced weight, improved equipment and connectivity and, for the first time, the option of four-cylinder engines and rear (as opposed to all) wheel drive.

2008: E70 models debut with new xDrive all-wheel-drive system and three-row seating.
2009: X5M with 555-hp, 5-liter twin-turbo V-8 added.
2010: Exterior facelift, new engines, and eight-speed automatic added.
2013: Third-generation F15 model introduced. Range includes four-cylinder diesel, teamed with rear-wheel drive, as well new six- and eight-cylinder units, plus lightened xDrive system; multimode Drive Dynamic Control programs handle ride, handling, and off-road behavior. Concept X5 eDrive presented at Frankfurt show with plug-in hybrid drive using a four-cylinder, twin-turbo gasoline engine and a 113-hp electric motor.
2014: X5 M debuts with 4.4-liter M TwinPower twin-scroll turbo V-8 engine giving 575 hp, the most power ever for a BMW four-wheel-drive model.
2015: X5 xDrive40e is first production BMW hybrid car, with total system power of 313 hp from its 2-liter gasoline engine and electric motor. Zero-emissions range is 60km and CO_2 emissions drop to 77g/km.

BMW showed its seriousness about China by launching its new CS Concept style statement in Shanghai. Though tipped as a future 8-Series, the production model was canceled because of the economic crisis. The concept's interior was exquisitely detailed.

of 2007, Reithofer revealed that the BMW board was putting the finishing touches to a fresh and highly ambitious strategy to take the group forward. The strategic review presented in September of that year would later become Strategy Number One, which gave a clue to its principal priority—to secure BMW's position as the world's leading provider of premium products and services for individual mobility. Unusually for an industry often dominated by pragmatic, short-term thinking, BMW's review set out objectives as far as thirteen years ahead, targeting two million vehicle sales by 2020 and setting further goals including an 8 to 10 percent return on sales by 2012; to achieve this, said Reithofer, efficiency savings and productivity improvements totaling €6 billion would be implemented over the coming five years.

Launched at the Frankfurt show in September 2007 were the Mini Clubman, a small station wagon with an unusual door arrangement, further EfficientDynamics models, and as a world first, a diesel engine in a luxury sports car—the 6-Series. Upcoming new models, said Reithofer, would include the X1 small SUV, the X6 sports activity coupe, and

an SUV from Mini, which would become the Countryman, and which would be built not in capacity-constrained Oxford but by Magna Steyr.

With BMW Motorrad having recently acquired Husqvarna Motorcycles, a famous once-Swedish brand producing motocross and off-road bikes in Italy, Reithofer said he was still open to further acquisitions or the creation of a fourth car brand. In the event, the closest BMW would later come to a fourth brand would be Project i, the pioneering skunk works, think tank, and channel for electric vehicles and sustainable transport launched in 2011.

The period following the strategic review was characterized less by new model launches and more by behind-the-scenes activity, such as the setting up of the secretive Project i in March 2008; this led to the announcement of the Mini E program and an extensive field trial of electric Minis with just one slight disadvantage: the whole back seat was replaced by the bulky battery pack. Major product actions in 2008 were limited to the launch of the X6, a new and controversial shape of Sports Activity Coupe, in January,

Mini body styles began to proliferate after 2007, beginning with the Clubman miniwagon featuring novel double rear doors and a half-sized "club" side door.

The Mini E field trial provided valuable electric-car feedback to inform the i-car program that was just gaining momentum. ❯

^ **BMW used the X6 Sports Activity Coupe to open up a further niche in the SUV segment.**

and the presentation of the fifth-generation 7-Series, with its design much moderated from its confrontational predecessor, in July. Shortly after that, however, came the Lehman Brothers collapse—and everything changed.

In an interview for this book, Ian Robertson, who was appointed BMW board member for global sales and marketing earlier that year, remembered the position very clearly:

First of all, the financial crisis was largely a surprise for everybody and for every industry. Perhaps the most surprising thing was that the whole world moved in a negative direction at the same time. [At BMW] we had a number of advantages. We had started to develop Strategy Number One the year before, so we had been thinking of a number of scenarios, and volatility was part of our thinking. We had already preplanned some "what-ifs." But I think the most important thing we did was to react very quickly to a set of circumstances that were difficult to read and draw conclusions from.

I can remember a very important discussion we had at the Paris auto show in October 2008,

BMW X6 (E71), 2008–2014, and BMW X6 (F16), 2014–Present

Causing something of stir when it debuted in early 2008, the X6 (E71) was claimed by BMW to be the pioneer of an all-new Sports Activity Coupe market segment for buyers who wanted a 4×4's high driving position as well as the responsiveness of a sports car. Using the X5's platform and engines, the X6 had a lower roofline, fastback tailgate, and large wheels in swollen arches for a powerful macho effect. The all-turbo launch lineup was the first in BMW's history.

2007: X6 Concept Car unveiled at Frankfurt show, with bold, high-riding coupe style.
2008: Production models launched with a choice of three six-cylinder engines and the twin-turbo, 4.4-liter V-8 xDrive50i with 408 hp. All models have full-time AWD, EfficientDynamics, and Dynamic Performance Control to distribute torque laterally as well as longitudinally within the xDrive system.
2009: X6M added, with 555-hp version of 4.4-liter V-8 engine and lowered suspension. Transmission and Dynamic Performance Control recalibrated for sportier driving. Concept X6 ActiveHybrid presented, using the V-8 engine and two-mode transmission co-developed with GM and Daimler.
2014: New-generation (F16) models replace original range, with engine choices and upgrades paralleling those of the X5.
2015: X6M, with 575 hp and 280km/h top speed in de-restricted form.

Reflecting the influence of chief designer Karim Habib, the fifth generation of BMW's flagship luxury sedan presented a smoother, less confrontational appearance. In order to meet new pedestrian safety rules, the hood line was raised, with larger BMW grilles and deeper headlights; the trunk line was gentler, and the sumptuous interior returned to the earlier integrated dashboard style. The all-new chassis, with double-wishbone front suspension and the new integral V axle at the rear, included the option of Integral Active Steering with speed-related rear-wheel steer. Six-speed automatic transmissions with electronic selectors complemented a new range of EfficientDynamics engines, and improved iDrive and comprehensive active safety systems such as Night Vision were included.

2008: New 7-Series launched at Paris show with choice of gasoline and diesel straight-six engines and novel V-8 on 750i incorporating twin turbochargers between the cylinder banks for compactness and quick warmup. Long-wheelbase versions also offered. Concept 7-Series ActiveHybrid displays V-8 gasoline-electric powertrain with electric motor integrated into new eight-speed automatic transmission.

2009: 760i and 760iL with new all-aluminum twin-turbo V-12 engine with 544 hp introduced, also featuring eight-speed automatic transmission. Higher-powered 740d added, with 306 hp; 750i and 750iL models gain option of xDrive AWD system.

2010: Equipment now includes improved Night Vision and in-car Internet as well as real-time traffic information for navigation system and head-up display.

2012: Introduction of eight-speed automatic across the range; ActiveHybrid7 has 3-liter six-cylinder gasoline engine and synchronous electric motor for 158g/km CO_2 emissions.

*when I was looking at our monthly figures for
September that year. They were coming through on
my Blackberry, and they were minus 30, minus 35,
minus 40 percent—and it was everywhere. I said to
Norbert Reithofer, "There's something very serious
happening out there, and whilst I'm not sure how it
will develop, we need to review our forward plans."
The decisions we took over the next three weeks
were very much geared to protecting our position.
We decided to take seventy-eight thousand cars
out of the production schedule before Christmas,
and I can remember the serious and complex
discussions we had within the board meeting.
Our schedule was fixed and the suppliers were all
geared up. It was a hard debate but in the end we
decided we would do it.*

*We came out of that year with our cash flow in
much better shape than our competitors. Did we
know that the overall economic situation was going
to be as bad as it turned out to be? No, but we knew
we were seeing something and that we needed to
demonstrate quick thinking, speedy reactions and, in
many respects, take brave decisions.*

∧ **Sales director Ian Robertson (pictured) and CEO Norbert
Reithofer were quick to spot the signs of a downturn in
global sales.**

An important factor working in BMW's favor was that, earlier
in the decade, it had developed a manufacturing system
designed to be flexible on the downside, as well as the up; in
a crisis there is nothing worse than being stuck with too much
capacity, too much tied up in inventory, and too many fixed
overheads. Robertson explains: "We were able to reduce our
capacity fairly simply by going into the working-time account
with our employees. But nevertheless, that's a very difficult
decision to take: we had the flexibility, but this was exceptional.
However, it meant we largely managed to avoid too much
inventory, with the attendant cash-flow problems that brings."

As the sales returns for the following months came in,
the scale of the collapse became clearer. The first quarter of
2009 saw BMW revenues drop over 13 percent and deliveries
fall some 21 percent across the group; however, even that
figure underestimated the fall in production as, in order to
minimize stocks, BMW continued to deliver more vehicles
than it built.

By May 2009 the European car market was 14 percent
smaller than the year before, but Reithofer nevertheless
retained an optimistic outlook, pointing to BMW's strong
liquidity position, its continuing status as the world's leading
premium brand, and the fact that it had at last nudged ahead
of Lexus and Mercedes in the United States. Indeed, said
Reithofer, investment would continue in the near doubling of
capacity at the Spartanburg plant, and other new products,
such as the X1, the 5-Series GT, and the new Z4 sports car,
would help draw in new customers.

Proof of the wisdom of BMW's strategy came with
the publication of the company's annual report for 2009:
while the BMW Group took a 10.4 percent hit in terms of
vehicle sales and its automotive operations did lose money,
the group results remained positive overall and a dividend
was still paid to shareholders. Mercedes-Benz, on the other

Making its debut in 2008, the new F-01 7-Series avoided controversy with a cleaner, fresher look and a return to a smoothly integrated interior design.

Karim Habib, responsible for the 2008 7-Series, has been BMW chief designer since 2012.

210

hand, was over 15 percent down and incurred painful losses of €2.6 billion, the consequences of which would hamper it in the years that followed.

Despite the uncertain outlook for world markets, the BMW Group's portfolio of fresh and about-to-be-launched models gave grounds for optimism, and in an interview with the *Financial Times* at the Frankfurt show in September, Reithofer brushed away suggestions that BMW might need to team up with another industry player in order to assure its survival. "Size isn't the only thing that matters," he told the *FT*. "We think our strategy as a premium carmaker is the right one—not to mix up premium and volume brands."

Adamant that there was scope for BMW to survive without the protection of the extra volume from a mass-market manufacturer—precisely the mistake it made in the 1990s with the Rover acquisition—Reithofer said he thought the global car industry would recover fully only in 2010, remaining "absolutely convinced" that demand for luxury cars would bounce back to its pre-crisis levels.

Perhaps more importantly still, there was a palpable sense of excitement surrounding the models BMW displayed

at its large Frankfurt show pavilion. The undoubted star of the show was a dramatic concept car, the Vision EfficientDynamics—the inspiration behind the production i8 sports car. With its broad and low stance, its complex, flowing, multilayer bodywork, and its electric-blue highlights, it did indeed appear like a fast-forward vision of the future, and this was made all the more real by an imaginatively engineered powertrain providing startling performance from a three-cylinder, 1.5-liter engine, a clutch of electric motors, and a plug-in hybrid battery. Promising equally remarkable fuel economy, this to the 2009 show goer was a beguiling foretaste of the future that was hard to resist.

By comparison, the showroom-ready offerings came across as somewhat mundane. The X1, though commercially important, lacked distinction, especially from its larger X3 brother; the 5-Series GT, built on the platform of the larger 7-Series launched the year before, appeared an ungainly answer to BMW's self-imposed task of combining the attributes of station wagon, sedan, limousine, and high-riding people carrier in a single vehicle. On the technical side, however, industry watchers quickly spotted the eight-speed

▲ The battery-powered i3 was shown in several different concept iterations before appearing as a production car in 2013.

▼ The Vision EfficientDynamics concept of 2009 marked the start of a new era in sophisticated, ecologically conscious sports cars.

^ **The 5-Series GT sought to combine the attributes of a sedan, wagon, and people carrier.**

automatic transmission—as well as a hybrid derivative—on the 7-Series. Manufactured by ZF, this transmission helped trigger a ratio race among carmakers that would see everyday vehicles with nine, ten, or even more speeds.

We have given EfficientDynamics a face, Reithofer told shareholders in May 2010. "The message is: Sustainable mobility can be at least as exciting as conventional mobility—if you get the best engineers and the best designers together to develop the best result." Again, the confidence was well justified, for on the strength of its low-CO_2 EfficientDynamics-equipped vehicle fleet, BMW had been selected as the official mobility partner for the high-profile 2012 London Olympics, and all the financial indicators were pointing in the right direction: sales were up 14 percent in the first quarter, the new 5-Series had been launched to great acclaim, and the Mini Countryman was proving a big hit with customers.

Like its bigger 7-Series companion, the new 5-Series marked a climbdown from the intricate style of the Bangle-era E60 model; many of the 7-Series' new technologies, such as four-wheel steering and eight-speed automatics, now made their first appearance on the smaller car, and on the manufacturing side the move to a more module-based assembly practice anticipated the much stricter 35-up modular approach that would be introduced with the sixth-generation 7-Series in 2015.

BMW X1 (E84), 2009–2015, and X1 (F16), 2015–Present

The long-expected compact sports activity vehicle from BMW had been previewed by a close-to-market concept at the Paris show in 2008, so the very X3-like styling of the series production X1 was little surprise when the definitive model appeared at the Frankfurt show the following autumn. Based on the engineering of the 1-Series hatchback, the X1 had a selection of three gasoline and three diesel engines, as well as the choice of rear-wheel drive (sDrive) or all-wheel drive (xDrive). With EfficientDynamics technologies, the X1 was the first X model to feature engine stop-start. In 2015 the X1 moved onto an all-new front-drive platform.

2008: Concept X1 at Paris show closely echoes X3 styling and is just 100mm shorter.

2009: Production X1 launched at Frankfurt show, with 2- and 3-liter gasoline engines and three different 2-liter diesels up to 204 hp in power.

2010: Top-of-the-range X1 xDrive23d now available.

2011: New X1 xDrive20d EfficientDynamics edition is the most economical BMW X model to date, with CO_2 emissions of 119g/km.

2012: Revisions to styling and interior made, and option of eight-speed automatic transmission introduced.

2015: End of production, with 730,000 sold globally. Replaced by all-new F48 series, based on front-wheel-drive platform shared with new 2-Series. All engines are transversely mounted 2-liter four cylinder units; the 192-hp xDrive20i gasoline is paired with eight-speed automatic transmission and AWD; diesels are 231 hp, 190 hp and 150 hp, the latter also available with front-only sDrive. All models include EfficientDynamics technologies, iDrive, navigation, and automatic tailgate opening.

▲ BMW provided a fleet of low-emission vehicles for the 2012 London Olympics.

Mini took a step up in size with the Countryman in 2010. The first to be offered with four-wheel drive, it was
▼ built by Steyr in Austria.

All this was to feed into a growing momentum behind BMW, Rolls-Royce, and Mini products, improved product development and manufacturing systems, and a sense that the group was operating at full throttle once more; motorcycles were growing almost 17 percent on the back of the revised high-performance K1300 inline four, and BMW announced further car plant expansions in China and at the US Spartanburg facility. This would now become an X-series center of excellence following the switching of the X3 from Steyr in Austria to the US location.

But it was two other announcements that snatched the lion's share of the headlines: BMW teaming up with PSA Peugeot Citroën on front-drive hybrid technologies and, more controversially, the confirmation—feared by many enthusiasts—that BMW was developing a front-wheel-drive architecture that would be shared by Minis and smaller BMW models below the 3-Series. In response to protestations from certain commentators and many enthusiast magazines, who feared the switch to front-drive would lose BMW its unique appeal in the compact car market, Reithofer presented the results of a survey of 1-Series owners. "Before getting started," he said, "we asked owners of BMW 1-Series cars: how important is a rear-wheel drive in your BMW? Contrary to the

Six generations of 3-Series, from top to bottom: F30 (2012), E90 (2004), E46 (1998), E36 (1990), E30 (1982), and E21 (1975).

BMW 5-Series (F-10), 2009–Present

By this, its sixth generation, the 5-Series had become a big seller, remaining the segment leader four years in succession and developing a loyal following worldwide. Replacing the Bangle-era E60 version in early 2010, the new F10 borrowed heavily from the engineering and the calmer, less confrontational styling of the 7-Series launched the year before. Initial engine choices ranged from the four-cylinder diesel 518d and 520d to twin-turbo diesel and gasoline sixes, with the V-8 being reserved for the top-performance M5. Something of an anomaly was the 5-Series Gran Turismo, a bulkier and taller five-door fastback based on the longer 7-Series platform and launched just ahead of the core sedan range.

2009: 5-Series Gran Turismo hatchback (F07) presented, with 3-litre diesel and gasoline engines or 407-hp V-8 gasoline, all featuring new eight-speed automatic.

2010: Launch of 5-Series sedan and long-wheelbase version for China.

2010: Touring station wagon version (F11) goes on sale; 535d and 525d engine options added, with 299 hp and 204 hp respectively.

2011: M5 launched with 4.4-liter, 560-hp twin-turbo V-8, seven-speed M double-clutch DriveLogic transmission, and new Active M differential; 520d EfficientDynamics edition combines 184-hp performance with 119g/km CO_2 emissions.

2012: ActiveHybrid 5 has 3-liter gasoline engine and 54-hp electric motor. Limited sale.

2013: M5 Competition Package offered, BMW's most powerful production car to date with 575 hp.

2014: Limited-edition "30 Jahre" M5 with 600 hp.

POWERING THROUGH THE DOWNTURN

Carrying on the fine tradition of elegant luxury BMW sports cars, the third generation of the 6-Series saw a refinement of the basic proportions laid down by the previous model, with a thrust-forward BMW grille, sweptback headlights, a crisper and higher waistline crease, and a more conventional rear deck, giving a longer and more elegant allure. As before, the chassis was drawn from 5-Series experience while the two turbo powertrains were 7-Series derived. Unusually, the Convertible was released first, and something of a surprise was the extension of the range with the Gran Coupe—in effect a four-door version of the regular 2+2 Coupe—in 2012.

2010: 6-Series Convertible (F12) introduced, with distinctive "fin"-style soft-top roof. Engines are the 320-hp turbo straight six (640i) and 407-hp V-8 with the twin turbochargers located within the V. Eight-speed automatic and EfficientDynamics systems are standard.

2011: Coupe versions added, and new 640d, with 313 hp, offers diesel alternative.

2012: M6 Coupe and Convertible announced. TwinPower Turbo V-8 has 560 hp for 0–100km/h acceleration in 4.2 seconds; seven-speed M DCT gearbox and Active M differential are standard and M Carbon Ceramic brakes optional.

2012: 6-Series Gran Coupe introduced, with 113mm-longer wheelbase, four-door coupe body style, low roofline, and frameless windows. New 450-hp V-8 engine with Valvetronic in 650i across the range.

2013: M6 Gran Coupe added.

2013: M6 Competition Package offered. Shared with the M5, this 575-hp V-8 is BMW's most powerful production car to date and its quickest-ever to 100km/h, at 4.1 seconds.

2015: Styling and engineering changes to whole lineup. Chassis uprated, with Dynamic Damper Control and Integral Active Steering (rear-wheel steer) and Active Drive roll stabilization.

▲ **Generation six of the 5-Series saw a retreat from the visual intensity of the previous design, and the 2012 3-Series would follow many of the same cues.**

larger vehicle classes, the vast majority of the 1-Series drivers responded that they were not concerned at all whether their car had a front- or rear-wheel drive, as long as it was a BMW. Customers trust us. And this is what matters."

Profits and Pressure to Perform

While BMW continued to reel off a string of record quarterly and annual results, proving conclusively that the financial crisis was fast disappearing in its rearview mirror, the investment community professed itself still not satisfied with BMW's profit margins, especially as Porsche had posted some startlingly good performances and new rival Audi was catching up at an impressive rate. Much effort thus went into improving processes and efficiencies to help streamline the industrial side of the equation, while growing commonality of engines and transmissions allowed for more sophisticated drivelines at competitive cost; all these improvements were a direct result of BMW's resolve to maintain its high level of R&D and vehicle-development expenditure even during the crisis years.

▲ Daimler CEO Dieter Zetsche was in full attack mode at the 2012 Geneva show, with the new A-Class aimed directly at BMW's 1-Series. *Daimler*

BMW 1-Series (F20), 2011–Present

Fears that the second-generation 1-Series would make the switch to front-wheel drive proved unfounded. The new F20 model based itself on rear-drive 3-Series engineering to retain what BMW insisted was a unique advantage in the market, and the evolutionary style stayed faithful to the distinctive rear-biased proportions of the original. A new-style grille folded gently over into the hood for a wider, fresher look, the rear lights now curled round onto the tailgate, and the interior ambience was upgraded. Most importantly, it was longer and wider, to the benefit of rear seat and trunk space, and the options list grew to take in the sophisticated systems once exclusive to larger BMW models.

2011: New 1-Series hatchback presented at Frankfurt show. Engines range from 116i gasoline to 120d diesel; transmissions are six-speed manual or eight-speed automatic.

2012: Three new powertrains introduced—116d EfficientDynamics, 125d and 125i, the latter both with 218 hp; the 116d ED is the most fuel-efficient BMW ever, with 99g/km CO_2 emissions. Three-door models (F21) and high-powered M135i added.

2012: 114i added as entry-level model; 114d Sports Hatch added, with 1.6-liter diesel engine giving 95 hp and 112g/km CO_2 emissions; xDrive all-wheel drive option added on 120d and M135i.

2013: Production of Coupe and Convertible, still based on older E82/88 architecture, ceases. New-generation Coupe (F22) and Convertible (F23) relaunched as 2-Series in 2014.

2015: Light facelift, with kidney grilles becoming wider and flatter, LED taillights and optional LED front lighting added, and navigation made standard. New twin-turbo gasoline engines range between 136 and 326 hp (M 135i); five diesel choices now include 1.5-liter three-cylinder 116d with 116 hp and 94g/km CO_2. Top-economy 116d EfficientDynamics Plus model gives 89g/km. Automatics gain a coasting function for further fuel saving.

Such was the pace of BMW's acceleration that in July 2011 it raised its full-year target to 1.6 million units; the result, when declared in January 2012, was 1.669 million, the best in BMW's ninety-four-year history. Rivals Mercedes-Benz (including Smart) and a fast-closing Audi could only manage 1.36 and 1.30 million units respectively. Ian Robertson, praising the performance of the X3, X1, and segment-leading 5-Series, said that he looked forward to the next generation of the 3-Series the following month.

Conscious of BMW's growing momentum, Mercedes-Benz CEO Dieter Zetsche declared "A for Attack" as he unveiled his company's new A-Class (aimed at the BMW 1-Series) at the 2012 Geneva show; earlier, he had announced a new model program of unprecedented intensity, with ten new ranges by 2020. Simultaneously, he set out goals of two million sales, a 10 percent margin, and premium-market leadership, also by 2020; this in turn triggered a bidding war, with Audi, too, signing up to being number one and selling over two million units annually—by the same date.

POWERING THROUGH THE DOWNTURN

BMW X3 (F25), 2010–Present

The second generation of BMW's successful medium Sports Activity Vehicle, the X3, was larger, more sophisticated, and more harmoniously styled than its predecessor, but it takes a trained eye to spot the differences at first glance. Under the skin, entirely new suspension was matched by familiar permutations of EfficientDynamics engines and transmissions and, as a world first, electromechanical Servotronic steering as standard. Luxury-sector innovations such as head-up display and iDrive characterized the much-upgraded interior.

2010: Launch lineup comprises one gasoline and three diesel engine versions, ranging from 143 to 258 hp.

2011: Engine choice expanded to include entry-level xDrive20i gasoline with 184 hp and six-cylinder twin-turbo gasoline xDrive35i (306 hp) and 35d diesel with 313 hp.

2012: Whole range becomes twin turbo, xDrive28i gasoline now with four cylinders and 245 hp; all models now fitted with Driving Experience Control to select drive modes; X3sDrive18d (143 hp) with rear-drive becomes entry model.

2014: Revised frontal styling reflecting new 3-Series, with headlights stretched inward toward reshaped grilles; new-generation 2-liter diesel with 190 hp.

Major development programs in the motorcycle division brought the super-luxurious six-cylinder K1600 (above) as well as an entry into the premium scooter market with the C600 Sport (right) and C650GT.

As an aside, *Automotive News Europe* reported in April 2013 that though BMW was the clear market leader, it was Mercedes that was the frontrunner using a metric called average revenue per unit (ARPU), a measure employed by financial analysts to compare the "premiumness" of different brands. In 2012, said the report, Mercedes had generated an average of €45,800 in revenue per unit, according to the Barclays Equity Research measure. "The result is impressive," said the paper, "because Mercedes' ARPU is watered down by the much lower retail prices of its Smart minicars. BMW's average revenue per unit was €41,350 and Audi was far behind at €37,500."

Announcing a series of best-ever quarterly performances, fueled by large surges in Chinese demand, Reithofer brought the long-term two million volume target forward from 2020 to 2016. In the event, however, that target would be reached earlier still, in 2014, following two further record years and the beginning of an ambitious expansion of niche models derived from the volume F30 3-Series architecture. Over the course of the next few years, this program would see a bewildering array of new and often overlapping derivatives. The 3-Series, which had begun to look very similar to the 5-Series and almost as large, now comprised three models: a sedan, Touring station wagon, and 3-Series GT—in effect a higher-riding hatchback. But now the top end of the range was relabeled 4-Series and took in the fresh Coupe and Convertible, the Gran Coupe (which mimicked the 6-Series and was easy to confuse with the 3 GT), and the X4, a slightly shrunken rendition of the X6 sports activity coupe.

The 3-Series is always the big one for BMW: the sixth generation, launched in 2011, did more than just sell in large volumes—it also established a fresh design language that would be gradually rolled out across other model lines too. Larger and roomier but also lighter than the previous model, the F30 introduced the new low and wide frontal look, with the headlamps drawn inwards to meet the kidney grilles. Mechanically, the aluminum front suspension was new and the rear five-link system revised, and Variable Sport steering and Adaptive M Sport suspension were listed as extras. The F30 is notable also for having spawned an additional range, all Coupe, Cabriolet, and Gran Turismo versions being designated 4-Series.

2011: 3-Series sedan (F30) unveiled at Frankfurt show; TwinPower Turbo gasoline engines include four-cylinder 328i with 245 hp and six-cylinder 335i with 306 hp; diesels comprise 320d EfficientDynamics with 163 hp, 320d with 184 hp, and 330d with 258 hp. Transmission choices are six-speed manual and eight-speed automatic.

2012: 316d, 318d, and 320i (184 hp) added to lineup. Touring wagon versions with 5-Series-like styling added, along with 2-liter gasoline 316i as entry model. xDrive four-wheel-drive models and ActiveHybrid 3 become available.

2013: New 3-Series Gran Turismo range launched, with longer wheelbase, taller body, and coupelike fastback rear design. Five engine choices offered, from 143 to 205 hp.

2013: 4-Series Coupe (F32) has dedicated body style similar to 6-Series, lower, sleeker, and sportier than 3-Series, while engines include one six-cylinder and two fours. Convertible versions also introduced, with three-section folding metal hardtop that can be raised or lowered in 20 seconds.

2014: M3 sedan and M4 Coupé introduced, with turbocharged 3-liter six-cylinder engine giving 431 hp; transmissions are six-speed manual or the latest seven-speed M DCT with DriveLogic and Launch Control, plus Active M Differential.

2014: 4-Series Gran Coupe added to range, combining the style of a coupe with the functionality of a four-door hatchback or wagon. The same length and width as the Coupe, it has a higher roofline extending further rearwards to provide increased passenger headroom.

2014: M4 Cabriolet joins lineup.

2015: Minor exterior facelift to 3-Series sedan, and Touring, adding LED daytime running-light strips linking each pair of round projector units. New modular engine range begins with three-cylinder 318i gasoline, with 136 hp from 1.5 liters; new 2-liter four-cylinder gasoline gives 184 hp (320i) or 252 hp (330i); new 3-liter six-cylinder gives 326 hp in 340i. Four-cylinder diesels range from 116 hp (316d) to 190 hp (320d). Eight-speed automatic transmissions now offer coasting mode. xDrive now optional on 320i, 320d, and 330d and standard on 335d; plug-in hybrid 330e, with CO_2 emissions of only 49g/km, replaces ActiveHybrid3.

2016: 3-Series 325d with new four-cylinder diesel.

The two ranges were united by a new frontal identity set to distinguish almost all upcoming core BMW model lines. "The F30 3-Series was a very important step for us," said design director Karim Habib in an interview for this book. "We have been able to achieve a very low front, and we've made the important step of connecting the kidneys of the grille to the headlamps. That's something you will see a lot more of."

Even prior to this broadside assault on those midmarket niches, BMW had been working on many different levels towards its goal of sustainable solutions for personal mobility. The many initiatives aimed at securing the company's leadership position included the electric i3 megacity vehicle, announced in 2011 for a 2013 launch, hybrids for multiple model lines, including the 5- and 3-Series and the X5, and the rapturously received i8 plug-in hybrid sports car. Additionally, a hybrid and electric vehicle cooperation deal was signed with Toyota in 2011 and extended in 2013 to include fuel cells, "post-lithium" battery technologies, and joint sports-car development. Looking ahead, investments had already been made in schemes such as ParkatmyHouse (now JustPark) and the city car-share program DriveNow.

While specially prepared Minis had dominated the grueling Paris–Dakar Rally, back at Mini headquarters a further tranche of investment in the Oxford plant was earmarked to relieve chronic capacity constraints and prepare for the new generation of post-2014 Minis built on the UKL platform to be shared with upcoming BMW models, such as the 2-Series Active Tourer, the Grand Tourer, and, later still, the front-drive replacement for the F20 1-Series. And on the motorcycle side, business was being stoked up at both extremes of the lineup: the extravagant six-cylinder K1600 was a sensational entrant in the luxury class, and, after the failure of the C1 "scooter with a roof," built between 2000 and 2002, the company bravely decided to have another crack at the urban commuter market with the C600 Sport and C650GT premium scooters. These were given a much more positive reception than the ahead-of-its-time C1.

By 2013 BMW found itself increasingly capacity-constrained in Europe, its four plants working at 105 or 110 percent utilization; early in 2014 the decision was taken to outsource some Mini hatchback volume to the idle NedCar plant in Holland. So by the time of his final presentation to shareholders in May 2015, Norbert Reithofer was able to look back on a highly successful nine-year tenure at the top.

BMW X4 (F26), 2014–Present

Emboldened by the success of the larger X6, but also responding to the Audi Q5, BMW brought the midsize X4 to market in 2014. The company claimed the X4's muscular, high-riding stance combined the assertiveness of the X line with the elegance of a classical coupe. Longer but lower than an X3, the new model had a strong and aggressive presence on the road, aided by large wheels and tires, and complex surfacing on the nose, hood, and bodysides. In terms of engineering, the X4 closely paralleled the X3, with three gasoline and three diesel engine choices, all linked to the xDrive AWD system.

2013: BMW Concept X4 midsized Sport Activity Coupe unveiled.
2014: Production X4 launched as xDrive20i, 28i, 35i, 20d, 30d, and 35d.
2016: High-performance X4 xDriveM40i added, with newly developed TwinPower Turbo straight six developing 360 hp.

∧ **The four-wheel-drive Mini Countryman provided the ideal basis for the fearsome Mini ALL-4 rally car, which won the grueling 9,000km Dakar four times in a row.**

Calendar 2014, the fifth record year in a row, saw sales surge well past their 2016 target of 2 million, with Mini well over 300,000 and motorcycles at almost 125,000. The big bets were already in play—the front-drive 2-Series for the family market, the i3 and i8 as role models for sustainable premium lifestyle vehicles, and the multi-city DriveNow scheme to introduce a broader and less moneyed clientele to the privilege of driving a Mini or BMW. By general agreement one of the most skillful and successful CEOs in BMW's history, Reithofer boosted sales from 1.4 million in 2005 to over 2.1 million in 2014, despite the intervention of worst economic crisis the world had known since the 1920s; in that period he increased revenue by more than 50 percent, and the nineteen models listed by the three car brands in 2006 had swelled to forty by the time he handed the reins over to Harald Krüger.

And if there were any incipient problems passed on to Harald Krüger in the handover, they were those common to the whole of the automotive industry: the downturn in China, the world's largest premium market and for long the source of substantial revenue streams; the threat posed by upstart premium electric carmaker and Wall Street darling Tesla; and the lingering question marks over the future of diesel as a fuel for clean-running and emissions-compliant vehicles worldwide.

19

In Shape for the
Future

It was late in 2007 when a highly subversive idea began to take root at BMW. In line with CEO Reithofer's recently instigated Number One strategy, the group was to adopt an even longer and broader perspective in its forward planning, expanding its vision to take in the wider context in which future models were to operate. Sustaining the group's number-one position beyond 2030 or 2040 would require a radical rethink not only of the automobile itself but also of the way it was built, the way it was made available to the customer, and the type of environments it would be driven in.

The sheer scale of these tasks prompted the setting up of a select project group, akin to a skunk works, to begin looking afresh at the whole value chain for a new type of car and a new model of ownership. The first external indication of this group's activities came in summer 2008 with the revelation that an electric version of the Mini was under development and that five hundred pilot-build cars would be field trialed in the United States. Two years later came the BMW ActiveE, a battery-powered version of the 1-Series Coupe, again aimed at amassing experience in field trials, this time in Europe and China as well. By the spring of 2011 the skunk works had gained sufficient traction that BMW went public, turning the think tank into a new subdivision called BMW i and charging it with developing sustainable and pioneering mobility concepts.

For many months beforehand, BMW executives had been speaking with some concern about the rise of huge cities as the predominant human habitat: these large-scale urbanized

areas would require an entirely new type of car, described by BMW as a "megacity" vehicle. This vehicle would have to produce zero emissions, engineers explained, and in order to deliver the agile performance and handling demanded by BMW customers, it would have to have a lightweight carbon-fiber structure to compensate for the added weight of the batteries. With the launch of the i division, the megacity vehicle was named as the i3, and there was also to be a sports

∧ BMW's investment in Project i is the biggest in the
company's history. It includes not only the i3 pure-
electric car but also the i8 plug-in hybrid sports car,
carbon-fiber structures, new manufacturing systems,
< and fresh ideas on retailing, ownership, and car sharing.

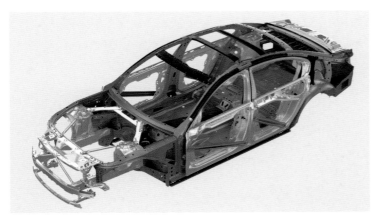

The 2015 7-Series was the first volume-built car to employ a carbon-fiber safety cell; further innovations included remote parking, gesture control, and the option of a plug-in hybrid powertrain (bottom). ⌄

car based on the groundbreaking Vision EfficientDynamics concept of 2009.

"It all derived from the Number One strategy," said BMW i division vice president Henrik Wenders in an interview for this book. "Of course the i project was controversial: we were talking about the largest investment ever made by BMW—it was a revolution. We were talking about a completely new vehicle architecture, the industrialization of the new lightweight material carbon fiber, a new plant in Leipzig. We were investing in a technology which didn't exist before, we were investing in new business models—all iconic changes."

"If you want to make yourself futureproof, you have to have a clear vision and a clear understanding of what is going to happen in the next decades," added Wenders. "The entire board, and also the supervisory board, shared our view of the main influences likely to affect future mobility in terms of regulation and energy supply, and agreed that sustainability will become a crucial factor when it comes to the purchasing behavior of our customers."

Following the presentation of evolving concept versions at a succession of motor shows, the final production-ready i3 was unveiled in July 2013. With rear-wheel drive and quick steering, the i3 was claimed to be a perfect fit with BMW's fun-to-drive mantra. The declared range was 160km, and for customers still concerned about running out of electricity, a range-extender version—complete with two-cylinder gasoline engine alongside the e-motor under the trunk floor—gave a near doubling of mileage. Along with the i3 came a major upgrade to BMW's ConnectedDrive networking service, giving energy management the top priority in the navigation system. Throughout, BMW made it clear that this was a thoroughly premium approach to sustainable mobility.

That much was made clear a year later with the launch of the i8 sports car. Priced well into six-figure territory, whether in euros, pounds, or dollars, the i8 immediately became BMW's most expensive and most prestigious model. It was exotic not only in its swoopy styling and upward-opening doors but also in its engineering: people could only marvel at the complexity of the gasoline-electric powertrain, the front e-motor and transmission, and the intelligent management of energy that promised remarkable feats of economy. Few seemed to worry that its main power unit was the smallest in the BMW catalog, at 1.5 liters; the press reports were glowing, and the new model was duly enshrined as BMW's flagship and green-technology ambassador.

Though the i3 and i8 were never expected to be big sellers, the new engineering thinking developed under the i project was quick to benefit other more mainstream BMW model lines. Hybrid versions of the X5, 3-Series, and new 7-Series soon appeared, and the 7-Series itself was constructed around a light but strong Carbon-Core central safety cell using experience from the Life Module of the i3. The 7-Series was the first to be designed around BMW's new modular cluster architecture, known internally as "35-up" as it would serve all rear- and four-wheel-drive models from the 3- and 5-Series upwards. The 7 again represented a cautious evolution of BMW's style and featured several interior and connectivity innovations, most notably the industry-first gesture control to operate cabin systems.

Rather more important strategically was the introduction of the 2-Series Active Tourer and, soon afterwards, the seven-seater Gran Tourer. These represented a major departure for BMW on almost every level. Not only were they built on the front-wheel-drive UKL2 platform (now shared with Mini), but they were unashamedly family-oriented machines with no sporting pretensions and little but the twin-kidney grille to distinguish them as BMWs; the 2-Series Coupe and Convertible, however, were still rear-wheel drive and based on the current 1-Series. In early 2016 BMW caused considerable excitement with a brand-new M car: based on

BMW i3 (i01), 2013–Present

The i3, the result of BMW's ambitious mission to create a "megacity" vehicle to respond to anticipated future market conditions, will go down in history as the first ever premium city car with electric drive. Most of the engineering secrets of the car were in the public domain well before its summer 2013 unveiling: the Drive module, consisting of the carbon-fiber chassis, battery pack, and rear-mounted electric motor; and the Life module, the passenger-carrying safety superstructure, again in carbon. The high-riding i3 Concept at the 2011 Frankfurt show gave a good idea of the design themes for the final production car, which was unveiled in summer 2013 as a polarizing design, 4m in length and with more angular bodywork and lower rear doors in place of the transparent doors on the Concept. The loft-style interior drew widespread praise for its Scandinavian clarity and simplicity.

2013: Production i3 launched at Frankfurt show with 170-hp electric motor and 18.8kWh lithium-ion battery, giving 0–100km/h acceleration in 7.2 seconds and a maximum range of160km. Range-extender option places twin-cylinder motorcycle engine in rear to extend range to 300km.
2014: Production increased to meet higher demand and launch in United States; second production facility for carbon fiber announced.
2015: i3 joins DriveNow car-sharing fleets in London, Berlin, and other cities.

BMW i8 (i12), 2013–Present

Such was the futuristic allure of the Vision EfficientDynamics concept car in 2009 that BMW had no option but to follow through with a production version of the visually striking 2+2-seater plug-in hybrid supercar. First shown in September 2013, the i8 retained much of the sci-fi flamboyance of the concept, as well as its carbon-fiber structure, its hybrid configuration, and its electrically driven front axle, though the three-cylinder turbo engine switched from diesel to gasoline and the dual-clutch gearbox became a planetary automatic. Performance statistics such as 362 hp peak power, 0–100km/h in 4.4 seconds, and superlow 49g/km CO_2 emissions caused the world to wake up to the possibility of a responsible supercar, making the i8 a truly pivotal product.

2011: Presentation of i8 Concept at Frankfurt show as preview of 2014 production model.

2012: i8 Spyder Concept shown at Beijing auto show, with two seats and shorter wheelbase.

2013: BMW i8 plug-in hybrid sports car announced for limited-volume production.

2014: Start of production at Leipzig; new Laserlight headlights mark a world first.

2015: i8 drive unit voted International Engine of the Year; sales total of 5,456 for full year exceeds aggregate sales of all other brands' premium hybrid sports cars.

2016: BMW i Vision Future Interaction concept car at Las Vegas Consumer Electronics Show is an open-top i8 with multiple display and control innovations, including AirTouch sensors for gesture control and three drive modes—self drive, assisted drive, and automated driving.

the 2-Series Coupe, the M2 mounted a powerful twin-turbo, straight-six engine in a compact and fine-handling chassis, and commentators were quick to praise BMW for returning to the original M3's roots as a thrilling but still relatively affordable choice for the sports enthusiast.

Closely related to the Active Tourer was the second-generation X1 sports activity vehicle with either front- or four-wheel drive, though externally it closely resembled the outgoing model. Still to complete the family is the next-generation 1-Series, which will become front-drive on the UKL2 platform, and with November 2015's sharp-suited Concept Compact Sedan giving a preview of how the new model could be expected to look.

All the while, BMW's production network has been expanding to keep pace with rising demand. The group's two-million-plus cars and motorcycles are built in some thirty facilities in fourteen countries worldwide; of the cars, roughly half are made outside Germany, with the biggest factory in the whole network being Spartanburg in the United States, where all the X models apart from the X1 are produced. The first Brazilian-built BMW came off the line in 2014, Mexico is set to join the network in 2019, and the new G-series motorcycles will be assembled in India and Brazil.

∧ **The year 2015 also saw the debut of two further plug-in hybrids—the 225xe and the 330e.**

A longer wheelbase gives the 2-Series Gran Tourer seating for seven.

The battery-powered i3 is the center of ambitious BMW initiatives such as ChargeNow and DriveNow, a car-sharing service that invites customers to "find it, drive it, drop it."

BMW 2-Series Coupe and Convertible (F22/23), 2013–Present, and BMW 2-Series Active Tourer, Grand Tourer (F45/F46), 2014–Present

The 2-Series marks a very important turning point for BMW: the moment when, after decades of advocating rear-wheel drive, it allowed the first front-wheel-drive models to enter its portfolio. These models were family-oriented people carriers, the five-seat Active Tourer and the longer-wheelbase Gran Tourer seating seven, both based on the UKL2 platform shared with the larger UK Mini models. But the story is made confusing by an entirely separate strand of 2-Series models, namely the Coupe, Convertible, and M2; these are technically unrelated and use the classical rear-drive platform from the 1-Series.

2012: BMW Concept Active Tourer displayed at Paris show with front-wheel drive, new 1.5-liter twin-turbo three-cylinder engine, and eDrive plug-in hybrid system. "Outdoor" concept follows, with integrated bicycle storage.

2013: 2-Series Coupe presented as replacement for 1-Series; rear-wheel-drive platform is based on 1-Series architecture, and engines include 220i four-cylinder gasoline with 184 hp and M235i straight six giving 326 hp.

2014: 2-Series Active Tourer announced for autumn release, initially with three-cylinder gasoline and four-cylinder diesel (218i and 218d) engines and front-wheel drive. Innovative features include multifunctional load area, head-up display, automated city braking, and parking assistant. Added later are 220i, 216d, 220d, and 225i, the latter two with xDrive all-wheel drive.

2014: 2-Series Convertible revealed, based on rear-drive Coupe and with similar engine options plus 228i M Sport with 245 hp.

2015: New versions of Coupe include 218i, 218d, 225d, and 220d with xDrive all-wheel drive.

2015: 2-Series Gran Tourer seven-seater introduced, with body extended by 214mm and featuring sliding center seat row. Engine lineup includes three-and four-cylinder gasoline units plus four-cylinder diesels giving up to 190 hp in Gran Tourer 220d xDrive.

2016: Launch of 225xe plug-in hybrid version of Active Tourer, first shown the previous September in Frankfurt. Twin-turbo 1.5-liter gasoline engine drives front wheels through eight-speed automatic, while 88-hp electric motor on rear axle gives all-wheel drive as well as pure e-drive for city use, with a range of up to 41km.

2016: World debut of new M2 Coupe, with M TwinPower Turbo 3-liter straight six giving 370 hp, linked to six-speed transmission and Active M Differential; seven-speed M Double Clutch gearbox is optional. Agile rear-drive chassis, M3-derived aluminum suspension, and thrilling performance are praised in media reports as reminiscent of the M2's mentors, the E30 M3 and the 2002 Turbo.

IN SHAPE FOR THE FUTURE

With some 2.247 million premium cars sold over the course of 2015, BMW is comfortably ahead of all its competitors, both old and new. But its longstanding competitor Mercedes-Benz is never to be underestimated and is rallying strongly with a vast range of models in every niche of the market; Audi is equally ambitious and has the advantage of the huge Volkswagen Group behind it, while Japan-based challengers Lexus and Infiniti are building ever stronger brand identities and are well respected in Asia and North America. And the powerful Korean group Hyundai is now doing what Lexus did in the 1980s and launching its own premium brand, Genesis. Long term, all of these brands could emerge as serious competitors.

BMW may be number one, but there cannot be any sense of complacency. Major challenges will confront the whole auto industry in the decades to come. Achieving true sustainability not just in vehicle emissions but also in manufacturing, sales and service is vital; so too will be smoothing the transition to electrified vehicles industrially as well as on a consumer level. But perhaps most crucially for a brand whose core message is driver enjoyment, the advance

∧ **Just to prove that it was still in touch with its traditional petrol-head fan base, in early 2016 BMW introduced the V-12-powered M760iLxDrive, the first-ever M car in the 7-Series, and the hotshot M2, which recalled the 2002 Turbo forty years earlier.**

^ **The BMW Welt brand center, close to the Munich factory, hosts regular exhibitions and concerts and is home to permanent displays of all six of the group's brands; it also has an important parallel role for customers as a handover center. More than twenty million people have visited BMW Welt since it opened in 2007.**

of autonomous driving systems must still allow people to drive and enjoy their cars, rather than simply letting the cars drive the people. And throughout, just as it has already done with ideas like EfficientDynamics and the creation of its i division, BMW must keep inventing and reinventing itself to retain its premium edge in the face of rivals hungry for a share of the cake.

That is the snapshot of the company as its first century draws to a close, and, just as in the hundred years just elapsed, the world will be expecting BMW to take a lead, to take the next bold step—the one that requires real vision.

BMW 7-Series (G11-12), 2015

The sixth generation of BMW's flagship 7-Series sedan debuted in the summer of 2015 after considerable pre-publicity surrounding its engineering, most notably the Carbon-Core center safety section claimed to be key in making the new model 130kg lighter. The external design, revealed in July, showed a more elegant evolution of previous themes, with less visual mass, a lower nose, and prominent chrome moldings around the side windows and above the rocker panel. Engineering innovations included new-generation six-cylinder engines and updated V-8s, together with the 740e plug-in hybrid with 49g/km CO_2 and a 40km electric range. Further technical highlights included air suspension, Integral Active Steering rear steer, and optional xDrive AWD as well as laser lights and remote-controlled parking operated from the new Display Key. The interior boasted multiple world-first innovations, including inductive charging for mobile devices, gesture control for audio and phone functions, and fifth-generation iDrive featuring touch-screen operation for the first time.

2015: 7-Series launched in standard- and long-wheelbase versions, with choice of 730d and 740i straight sixes, 750i V-8, and 740e plug-in hybrid. All feature eight-speed automatic transmission.
2016: M760Li xDrive launched as first M version of any 7-Series. New 6.6-liter TwinPower Turbo V-12 engine gives 610 hp for 0–100 km/h acceleration in a claimed 3.7 seconds. Optional M Driver package has speed limiter raised to 305km/h; eight-speed automatic recalibrated to allow sportier shifts when desired.
2016: Plug-in hybrid variants of BMW models given additional iPerformance designation to highlight technology transfer from i project models. 740e becomes 740e iPerformance.

20

The Next One Hundred Years

As the foregoing chapters will have made clear, BMW is something of a special case in the automotive world. It has gained immensely in authority and influence as the decades have elapsed, though its history has been impacted by significant crises, two of which could have proved fatal and a third—the Rover episode—that could have seen its independence jeopardized.

It was fifty-five years ago that the company found its true direction, cheekily challenging the stuffy premium establishment with its Neue Klasse sports sedan; now, BMW has itself become part of that establishment and therefore a target for others. Yet, as premium leader, it needs to remain a potent force for progress, not for preserving the status quo. Rather than holding out against inevitable change, BMW must continue to anticipate and embrace it, welcoming the opportunity to turn its culture of inventiveness to its own advantage.

So it is both significant and true to character that in marking its first one hundred years of enterprise, BMW chose to look forward rather than recall its past glories. The eagerly awaited Vision Next 100 concept car specially designed for the celebration avoids sentimental references towards back-catalogue products, apart from a double-kidney grille that hints at the 328 sports car first seen in 1936. Tellingly, that model was the real breakthrough design for the then-young carmaker, the product that brought BMW to the attention of the motorsport world and provided the launch pad for its sporting celebrity and its innovative engineering.

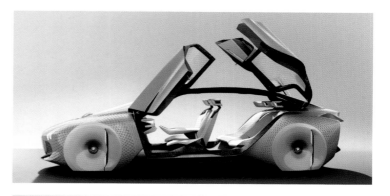

The profile and overhead views show the Vision Next 100's sleek configuration and generous space for passengers. BMW has not given any indication of how the vehicle will be powered, nor which wheels will be driven.

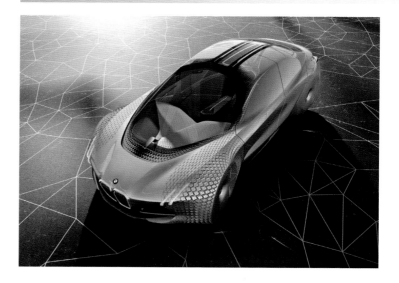

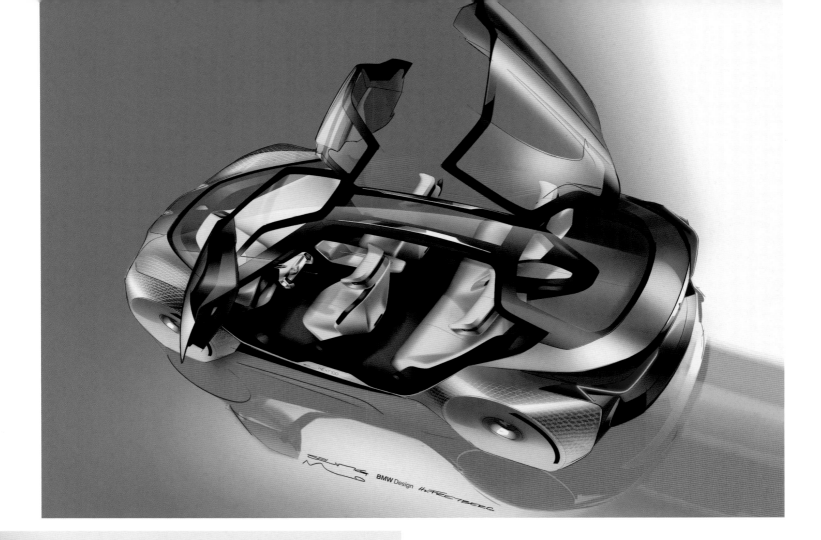

The prewar 328 was one of just eight key developments cited by CEO Harald Krüger in his centenary celebration address in Munich on March 7, 2016, a spectacular ceremony that took place one hundred years to the day after the formation of Bayerische Flugzeugwerke AG and just a kilometer away from the original workshops where the first BMW aero engines and motorcycles were made. Three qualities set BMW apart, declared Krüger: its capacity to learn and adapt, its aptitude for technical innovation, and its sense of responsibility towards society. "We have demonstrated on many occasions throughout our history that we are capable of learning fast and taking bold steps," he added. Appropriately, all three of these qualities are clearly evident in the Vision Next 100 concept, the distillation of BMW's design projections for twenty or thirty years hence. So, what does Vision Next 100 say about how BMW sees the future for itself and the industry?

Firstly, and most straightforwardly, Vision Next 100 may have a low-slung and aerodynamic nose, yet it is not a supercar but a four-seat sedan, even if those crowd-pleasing doors do open up dramatically skywards. The format, with the footprint of a 5-Series but the space of a 7-Series, places the concept at BMW's longstanding center of gravity, the premium sedan, and—also reassuringly—the sleek, minimalist dashboard houses a steering wheel of sorts, though nothing resembling conventional instruments or switches. And for good reason: so as to avoid information overload, only the most important messages are relayed to the display—and that display is the whole windshield, which acts as a screen for panoramic head-up projection.

The information highlighted on the screen depends on a further innovation: the option of the Ease or Boost mode of travel, a choice that even affects the configuration of the cabin itself. In Boost mode, the driver is fully in charge to enjoy driving the vehicle, as is the case today, and the display shows navigation directions and even recommended cornering lines and speeds. In Ease mode, the steering wheel retracts neatly into the dashboard, the center console swivels, and the car pilots itself autonomously, with the screen projections now showing entertainment and communications content as well

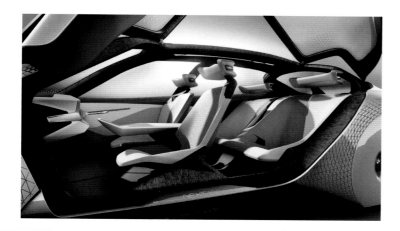

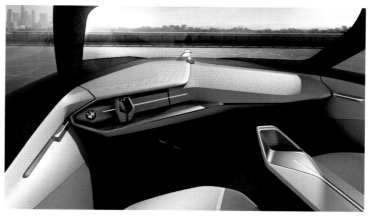

The Vision concept has the road footprint of today's 5-Series but offers the larger accommodation of a 7-Series. Rear seating is sculpted for two but access is easy.

BMW CEO Harald Krüger poses beside the Vision Next 100 concept car on the company's hundredth birthday in March 2016.

When driven in autonomous Ease mode (above), the concept's steering wheel retracts into the dashboard and the seats swivel outwards; in Boost mode (below), the driver is in charge, and hazards and recommended speeds and driving lines are projected onto the screen. The cabin layout changes, too, with seating and controls moving (bottom).

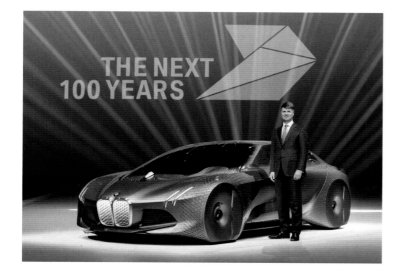

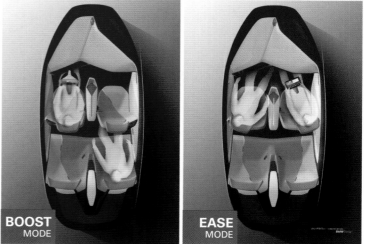

BOOST MODE

EASE MODE

as highlighting and explaining notable landmarks on the car's itinerary if desired.

Most dramatic is the exterior bodywork—sections of the body surface are able to change their shape and even their color depending on what the vehicle is doing and how it is being driven. When the steering is turned, the apparently rigid wheel arches stretch to keep covering the wheels regardless of their position; the color changes at the same time, with light shining through the hundreds of tiny polygons that change their shape to allow the whole surface to stretch. The aerodynamic benefits are likely to be considerable, as the body can always remain in the optimum configuration for the speed, and disruptive turbulence from the wheels and tires is all but

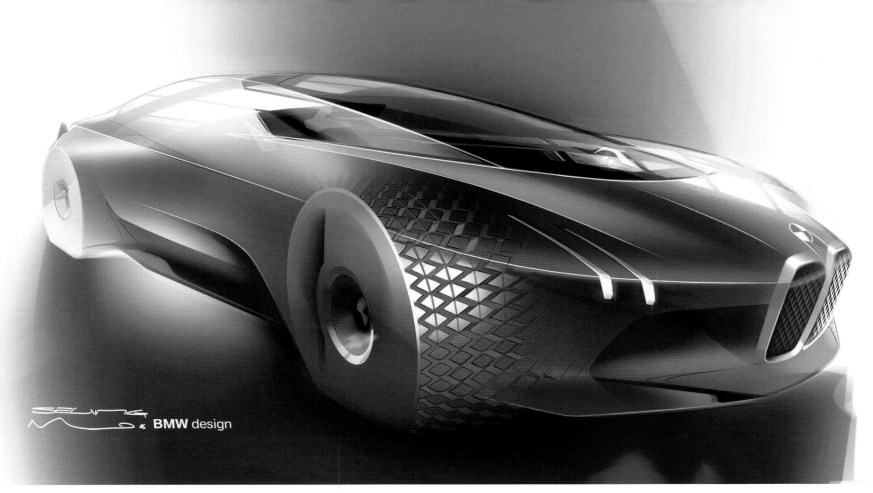

▲ Design sketch showing the visual effect generated by the Alive Geometry moveable body surfaces. As the steering is turned, the fender and wheelarch stretch to keep the wheel covered.

❯ As a compact four-passenger vehicle, the Next 100 sits at BMW's longstanding center of gravity, just as the pioneering 1500 sedan did over fifty years earlier.

▼ Close-up of the Alive Geometry surface: the tiny pyramids are visible in the driver's peripheral vision and flick up and change color to attract attention.

eliminated. BMW has given no detail about how this system operates, however, beyond saying it is a major challenge.

Chief designer Adrian von Hooydonk commented, "If you can imagine it, you have a strong chance of actually making it happen." Shape-shifting vehicles have been a dream of designers for decades, so perhaps this bold vision from BMW will come to be seen as the point where it all began.

Equally significant is what BMW does *not* tell us. Nothing, apart from the statement that it is zero emissions, is said about what powers the Vision concept. Why, when engines are so central to BMW culture and even figure in its name? The answer could be that the 2030–2040 time frame may be too far out for even BMW to have a firm idea, there may be several competing alternatives in the frame, or BMW may be exercising commercial caution by keeping any breakthrough technologies under wraps.

BMW i division vice president Henrik Wenders said in an interview for this book that by 2040 all of BMW's vehicle output would be electric in some form, with pure electrics, plug-in hybrids, conventional hybrids, and hydrogen models each making up around a quarter of total sales. In an address to shareholders in March 2016, Harald Krüger promised a total

of seven pure-electric or plug-in hybrid models by the end of the year, including an i3 with a more powerful battery that will give a range of 200km; in 2018 an "emotional brand shaper" i8 Roadster will be launched, and all future model lines will include iPerformance plug-in hybrid versions with technology transfer from Project i. There will be a plug-in hybrid Mini, too. In parallel, said Krüger, all BMW models will also be available with M-Performance packages, M-specification cars having quadrupled their sales since 2010.

Krüger's announcements came as he outlined a major new initiative of potentially greater significance than even EfficientDynamics or project i. Number One > Next is a future-oriented strategy designed to place BMW at the heart of the networking revolution and ensure that it continues to be a leading player as the whole automotive value stream turns itself inside out to focus on connectivity and user services rather than the hardware of the vehicle. This is a process BMW calls digitalization, and it takes in such developments as high-definition mapping, sensor technology, cloud technology, and even artificial intelligence. Under the umbrella of Number One > Next will come EfficientDynamics Next, to further

∨ **An earlier visualization of the concept, before the wheel enclosures had been added.**

improve drivelines; Project i 2.0, to ensure BMW a leading role in autonomous driving; i Ventures, to foster start-ups in the field of individual mobility; and iNext, an ambitious program to develop "a revolutionary new BMW i model that will raise premium individual mobility to previously unknown levels."

The iNext, set to appear early next decade, will be BMW's luxury flagship as well as the standard bearer for its next-generation technology—systems that can be expected to filter down to the broader base of the BMW ranges in the years that follow. Today's plug-in hybrids embody third-generation technology, with the fourth generation promising greater range and a fifth generation already under development; truly long zero-emissions range, up to 700km on a tank, is promised by the hydrogen fuel cell powertrains being co-developed with Toyota.

As an indication of the magnitude of the upheaval in prospect as car companies vie with technology giants such as Apple and Google for control of the revenue streams in the automotive sector, BMW is rethinking the way it develops its cars. Recognizing that communication will play a dominant role in the future connected, interactive, and automated world, BMW will begin to shape its cars from the inside out, starting with the interior and its myriad of integrated systems; interior design will become a separate discipline within vehicle development.

The biggest changes in the years to come are likely to be in interior design and operating systems. BMW has a history of innovation in this area, pioneering built-in navigation, iDrive, in-car WiFi hotspots, gesture control, AirTouch, and remote-control parking. The Vision Next 100 concept has provided a foretaste of the next steps, and in less than a decade the iNext production car will make these innovations a dynamic reality. The jump to the level of the ideas proposed in Vision Next 100 and delivered in iNext promises to be much greater than anything the auto industry has ever seen, and, as CEO Krüger is keen to assert, BMW is confident that transition and far-reaching change play to the company's strengths.

"Throughout its one-hundred-year history, the BMW Group has always reinvented itself," Krüger told the shareholders' meeting in March 2016. "As a pioneer of new technologies, the company has shaped change, within both the industry and the world of mobility. We are setting the standard with our Strategy Number One > Next, both now and in the future. We will lead the BMW Group into a new era, one in which we will transform and shape both individual mobility and the entire sector in a permanent way."

"The BMW Group is a company that has focused on the long term for the past one hundred years," he continued. "And this will continue to be our approach in the future. With our Strategy Number One > Next, we are looking ahead to the year 2020 and beyond that, up to 2025."

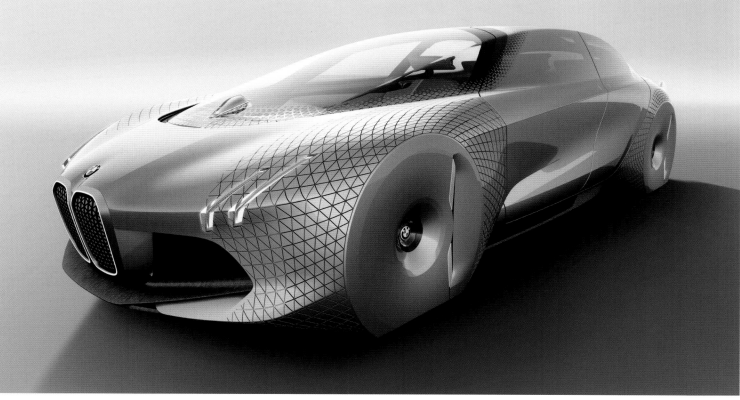

From the rear, the concept references today's L-shaped tail lights; moveable body surfaces can be seen on rear fenders and around rear screen. From the front, the new shape of the BMW double-kidney grille is clearly seen.

THE NEXT ONE HUNDRED YEARS

Index